Edward Albert Bowser

An Elementary Treatise on Hydromechanics, with Numerous Examples

Third Edition

Edward Albert Bowser

An Elementary Treatise on Hydromechanics, with Numerous Examples
Third Edition

ISBN/EAN: 9783744771122

Printed in Europe, USA, Canada, Australia, Japan

Cover: Foto ©berggeist007 / pixelio.de

More available books at **www.hansebooks.com**

BOWSER'S MATHEMATICS.

ACADEMIC ALGEBRA,
WITH NUMEROUS EXAMPLES.

COLLEGE ALGEBRA,
WITH NUMEROUS EXAMPLES.

AN ELEMENTARY TREATISE ON ANALYTIC GEOMETRY,
EMBRACING PLANE GEOMETRY,
AND AN
INTRODUCTION TO GEOMETRY OF THREE DIMENSIONS.

AN ELEMENTARY TREATISE ON THE DIFFERENTIAL AND INTEGRAL CALCULUS,
WITH NUMEROUS EXAMPLES.

AN ELEMENTARY TREATISE ON ANALYTIC MECHANICS,
WITH NUMEROUS EXAMPLES.

AN ELEMENTARY TREATISE ON HYDRO-MECHANICS,
WITH NUMEROUS EXAMPLES.

ELEMENTARY TREATISE

ON

HYDROMECHANICS,

WITH NUMEROUS EXAMPLES

BY

EDWARD A. BOWSER, LL. D.,

PROFESSOR OF MATHEMATICS AND ENGINEERING IN RUTGERS COLLEGE.

THIRD EDITION.

NEW YORK:

D. VAN NOSTRAND COMPANY,

23 MURRAY AND 27 WARREN STS.

1889.

Copyright, 1885,
By E. A. BOWSER.

Electrotyped by Smith & McDougal, 82 Beekman St., N.Y.

PREFACE.

THE present work on Hydromechanics is designed as a text-book for Scientific Schools and Colleges, and is prepared on the same general plan as the author's Analytic Mechanics, which it is intended to follow. Like the Analytic Mechanics, it involves the use of Analytic Geometry and the Calculus, though a geometric proof has been introduced wherever it seemed preferable.

The book is divided into two parts, namely, Hydrostatics and Hydrokinetics. The former is subdivided into three, and the latter into four chapters; and at the ends of the chapters a large number of examples is given, with a view to illustrate every part of the subject. Many of these examples were prepared specially for this work, and are practical questions in hydraulics, etc., taken from every-day life.

In writing this treatise, the aim has been to enunciate clearly the fundamental principles of the theory of Hydromechanics, to explain some of the most important applications of these principles, and to render more general the study of this interesting science, by presenting as simple a view of its principles as is consistent with scientific accuracy. Throughout the work a careful distinction has been made between those propositions which are necessarily true, being deduced from the definitions and axioms of the subject, and those results which are empirical.

In an elementary work of this kind there is not room for much that is new. I have drawn freely upon the writings of many of the best authors. The works to which I am principally indebted, and which are here named for convenience of reference by the student, are those of Besant, Lamb, Rankine, Boucharlat, Weisbach, Cotterill, Bland, Jamieson, Fanning, Pratt, Renwick, Stanley, Tate, Deschanel, Bossut, d'Aubuisson, Poncelet, Eytelwein, Prony, Starrow, Goodeve, Galbraith, Gregory, Twisden, Bartlett, Wood, Smith, Olmsted, Morin, Humphreys and Abbot, Fairbairn, Colyer, Barrow, and the Encyclopædia Britannica.

My thanks are again due to my friend and former pupil, Mr. R. W. Prentiss, of the Nautical Almanac Office, and formerly Fellow in Mathematics at the Johns Hopkins University, for reading the MS. and for valuable suggestions.

E. A. B.

RUTGERS COLLEGE,
NEW BRUNSWICK, N. J., April, 1885.

TABLE OF CONTENTS.

PART I.
HYDROSTATICS.

CHAPTER I.
EQUILIBRIUM AND PRESSURE OF FLUIDS.

ART.		PAGE
1.	Definitions—Hydrostatics, Hydrokinetics.	1
2.	Three States of Matter.	1
3.	A Perfect Fluid.	2
4.	Direction of Pressure.	3
5.	Solidifying a Fluid.	4
6.	Measure of the Pressure of Fluids.	4
7.	Pressure the Same in Every Direction.	5
8.	Equal Transmission of Fluid Pressure.	6
9.	Equilibrium of Pressures.	8
10.	Pressure of a Liquid at any Depth.	10
11.	Free Surface of a Liquid at Rest.	13
12.	Common Surface of Two Fluids.	15
13.	Two Fluids in a Bent Tube.	16
14.	Pressure on Planes.	17
15.	The Whole Pressure.	18
16.	Centre of Pressure.	21
17.	Embankments.	27
18.	Embankment when the Face on the Water Side is Vertical.	27
19.	Embankment when the Face on the Water Side is Slanting.	29
20.	Pressure upon Both Sides of a Surface.	33
21.	Rotating Liquid.	35
22.	Pressure at any Point of a Rotating Liquid.	37
23.	Strength of Pipes and Boilers.	39
	Examples.	42

CHAPTER II.

EQUILIBRIUM OF FLOATING BODIES—SPECIFIC GRAVITY.

ART.   PAGE

24. Upward Pressure, Buoyant Effort............................ 50
25. Conditions of Equilibrium of an Immersed Solid............ 52
26. Depth of Flotation 54
27. Stability of Equilibrium................................... 57
28. Position of the Metacentre; Measure of Stability........... 60
29. Specific Gravity... 66
30. The Standard Temperature................................... 67
31. Methods of Finding Specific Gravity........................ 69
32. Specific Gravity of a Solid Broken into Fragments.......... 72
33. Specific Gravity of Air.................................... 73
34. Specific Gravity of a Mixture.............................. 73
35. Weights of the Components of a Mechanical Mixture.......... 75
36. The Hydrostatic Balance.................................... 76
37. The Common Hydrometer...................................... 77
38. Sikes's Hydrometer... 78
39. Nicholson's Hydrometer..................................... 79
    Examples.. 80

CHAPTER III.

EQUILIBRIUM AND PRESSURE OF GASES—ELASTIC FLUIDS.

40. Elasticity of Gases.. 88
41. Pressure of the Atmosphere................................. 89
42. Weight of the Air.. 90
43. The Barometer.. 91
44. The Mean Barometric Height................................. 92
45. The Water-Barometer 92
46. Manometers 93
47. The Atmospheric Pressure on a Square Inch.................. 94
48. Boyle and Mariotte's Law................................... 95
49. Effect of Heat on Gases.................................... 98
50. Thermometers—Fahrenheit, Centigrade, Reaumur............... 99
51. Comparison of the Scales of these Thermometers............ 100
52. Expansion of Mercury...................................... 101
53. Dalton's and Gay-Lussac's Law............................. 101
54. Pressure, Temperature, and Density........................ 103
55. Absolute Temperature...................................... 105

ART.	PAGE
56. The Pressure of a Mixture of Gases	106
57. Mixture of Equal Volumes of Gases	107
58. Mixture of Unequal Volumes of Gases	108
59. Vapors, Gases	108
60. Formation of Vapor, Saturation	109
61. Volume of Atmospheric Air without its Vapor	110
62. Change of Volume and Temperature	110
63. Formation of Dew—the Dew Point	112
64. Pressure of Vapor in the Air	112
65. Effect of Compression or Dilatation on Temperature	113
66. Expansion of Bodies—Maximum Density of Water	113
67. Thermal Capacity—Unit of Heat—Specific Heat	115
68. Specific Heat at a Constant Pressure, and at a Constant Volume	116
69. Sudden Compression of a Mass of Air	118
70. Mass of the Earth's Atmosphere	120
71. The Height of the Homogeneous Atmosphere	120
72. Necessary Limit to the Height of the Atmosphere	121
73. Decrease of Density of the Atmosphere	122
74. Heights Determined by the Barometer	124
Table of Specific Gravities	130
Examples	131

PART II.
HYDROKINETICS.

CHAPTER I.

MOTION OF LIQUIDS—EFFLUX—RESISTANCE AND WORK OF LIQUIDS.

75. Velocity of a Liquid in Pipes	136
76. Velocity of Efflux	137
77. The Horizontal Range	140
78. Time of Discharge when the Height is Constant	141

viii   CONTENTS.

ART.		PAGE
79.	Time of Emptying any Vessel	142
80.	Time of Emptying a Cylinder into a Vacuum	144
81.	Time of Emptying a Paraboloid	145
82.	Cylindrical Vessel with Two Small Orifices	145
83.	Orifice in the Side of a Conical Vessel	146
84.	Velocity of Efflux through an Orifice in the Bottom	147
85.	Rectangular Orifice in the Side of a Vessel	149
86.	Triangular Orifice in the Side of a Vessel	151
87.	Time of Emptying any Vessel through a Vertical Orifice	156
88.	Efflux from a Vessel in Motion	158
89.	Efflux from a Rotating Vessel	160
90.	The Clepsydra, or Water-Clock	161
91.	The Vena Contracta	162
92.	Coefficient of Contraction	163
93.	Coefficient of Velocity	164
94.	Coefficient of Efflux	164
95.	Efflux through Short Tubes, or Ajutages	165
96.	Coefficient of Resistance	167
97.	Resistance and Pressure of Fluids	170
98.	Work and Pressure of a Stream of Water	172
99.	Impact against any Surface of Revolution	174
100.	Oblique Impact	178
101.	Maximum Work done by the Impulse	180
	Examples	181

CHAPTER II.

MOTION OF WATER IN PIPES AND OPEN CHANNELS.

102.	Resistance of Friction	185
103.	Motion of Water in Pipes	186
104.	Uniform Pipe connecting Two Reservoirs	188
105.	Coefficient of Friction for Pipes	191
106.	The Quantity Discharged from Pipes	194
107.	The Diameter of Pipes	197
108.	Sudden Enlargement of Section	199
109.	Sudden Contraction of Section	201
110.	Elbows	204
111.	Bends	206
111a.	Equivalent Pipes	207
111b.	Discharge Diminishing Uniformly	208

CONTENTS. ix

ART.		PAGE
112.	General Formula for all the Resistances	209
113.	Flow of Water in Rivers and Canals	211
114.	Different Velocities in a Cross-Section	212
115.	Transverse Section of the Stream	215
116.	Mean Velocity	216
117.	Ratio of Mean to Greatest Surface Velocity	216
118.	Processes for Gauging Streams	218
119.	Most Economical Form of Transverse Section	221
120.	Trapezoidal Section of Least Resistance	222
121.	Uniform Motion	224
122.	Coefficients of Friction	226
123.	Variable Motion	228
124.	Bottom Velocity at which Scour Commences	232
125.	Transporting Power of Water	233
126.	Back Water	235
127.	River Bends	236
	Examples	237

CHAPTER III.

MOTION OF ELASTIC FLUIDS.

128.	Work of the Expansion of Air	242
129.	Velocity of Efflux of Air according to Mariotte's Law	244
130.	Efflux of Moving Air	247
131.	Coefficient of Efflux	249
132.	The Quantity Discharged	250
133.	Coefficient of Friction of Air	251
134.	Motion of Air in Long Pipes	252
135.	Law of the Expansion of Steam	254
136.	Work of Expansion of Steam	256
137.	Work of Steam at Efflux	257
138.	Work of Steam in the Expansive Engine	259
	Examples	260

CHAPTER IV.

HYDROSTATIC AND HYDRAULIC MACHINES.

139.	Definitions	263
140.	The Hydrostatic Bellows	263

ART.		PAGE
141.	The Siphon	264
142.	The Diving Bell	266
143.	The Common Pump (Suction Pump)	268
144.	Tension of the Piston-Rod	270
145.	Height through which Water Rises in One Stroke	271
146.	The Lifting Pump	273
147.	The Forcing Pump	274
148.	The Fire Engine	276
149.	Bramah's Press	276
150.	Hawksbee's Air-Pump	277
151.	Smeaton's Air-Pump	279
152.	The Hydraulic Ram	280
153.	Work of Water Wheels	282
154.	Work of Overshot Wheels	283
155.	Work of Breast Wheels	284
156.	Work of Undershot Wheels	285
157.	Work of the Poncelet Water Wheel	286
158.	The Reaction Wheel; Barker's Mill	288
159.	The Centrifugal Pump	290
160.	Turbines	292
	Examples	294

HYDROMECHANICS.

PART I.
HYDROSTATICS.

CHAPTER I.

EQUILIBRIUM AND PRESSURE OF FLUIDS.

1. Definitions.—Hydromechanics is the science which treats of the equilibrium and motion of fluids. It is accordingly divided into two parts, *Hydrostatics* and *Hydrokinetics*.

Hydrostatics treats of the equilibrium of fluids.

Hydrokinetics treats of the motion of fluids.

The object of the science of *Hydrostatics* is to determine the equilibrium and pressure of fluids, the nature of the action which fluids exert upon one another and upon bodies with which they are in contact, and the weight and pressure of solids immersed in them, and to explain and classify, under general laws, the different phenomena to which they give rise.

2. Three States of Matter.—Bodies exist in three different states, depending upon the manner in which their particles are held together. They are either *solid* or *fluid*; and the latter are either *liquid* or *gaseous*.

Solid bodies are those whose particles are held together so firmly that a certain force is necessary to change their forms or to produce a separation of their particles. If a solid be reduced to the finest powder, still each grain of the powder is a solid body, and its particles are held together in a determinate shape.

Fluids are bodies, the position of whose particles in reference to one another is changed by the smallest force. The distinguishing property of a fluid is the perfect facility with which its particles move among one another, and as a consequence its readiness to change its form under the influence of the slightest effort.

Fluids are of two kinds, *liquids* and *gases*. In a *liquid* there is a perceptible cohesion among its particles; but in a *gas* the particles mutually repel one another. Every *solid* body possesses a peculiar form of its own, and a definite volume; *liquids* have only a definite volume, but no peculiar form; and *gases* have neither one nor the other. If a *liquid*, such as water, be poured into a tumbler, it will lie at the bottom, and will be separated by a distinct surface from the air above it; but if ever so small a quantity of *gas* be introduced into an empty and closed vessel, it will immediately expand so as to fill the whole vessel, and will exert some amount of pressure upon the interior surface.

3. A Perfect Fluid.—Fluids differ from each other in the degree of cohesion of their particles, and the facility with which they will yield to the action of a force. Many bodies which are met with in nature, such as water, mercury, air, etc., possess the properties of fluids in an eminent degree, while others, such as oil, tallow, the sirups, etc., possess a less degree of fluidity. The former are called *perfect fluids*, and the latter *viscous* or *imperfect fluids*. In this work, only perfect fluids will be considered.

A perfect fluid is an aggregation of particles which yield at once to the slightest effort made to separate them from one another.

Fluids are divided into two classes, *incompressible* and *compressible*. The former are sometimes called *inelastic* and the latter *elastic* fluids.

Incompressible fluids are those which retain the same volume under a variable pressure. *Compressible fluids* are those in which the volume is diminished as the pressure upon it is increased, and increased as the pressure upon it is diminished.

The term incompressible cannot strictly be applied to any body in nature, all being more or less compressible. But on account of the enormous power required to change, in any sensible degree, the volumes of *liquids*, they are treated in most of the researches in hydrostatics as incompressible or inelastic fluids. It was shown by Canton, in 1761, that water under a pressure of one atmosphere, *i. e.*, of about one ton on each square foot of surface, undergoes a diminution of forty-four millionths of its total volume.* All *liquids* are therefore regarded as incompressible. Water, mercury, wine, etc., are generally ranged under this class. The *gases* are highly compressible, such as air and the different vapors.

4. The Direction of the Pressure of a Fluid on a Surface.—If an indefinitely thin plate be made to divide a fluid in any direction, no resistance will be offered to the motion of the plate in the direction of its plane, *i. e.*, there will be no tangential resistance of the nature of friction, such, for instance, as would be exerted if the plate were pushed between two flat boards held close to each other. Hence the following fundamental property of a fluid is obtained from its definition :

The pressure of a fluid is always normal to any surface with which it is in contact.

---

* Galbraith's Hydrostatics; Gregory's Hydrostatics.
The compressibility of water per atmosphere at 8° C., as given in Everett's *Units and Physical Constants*, is 48.1 millionths. Ency. Brit., Vol. XII, p. 439.

5. Solidifying a Fluid.—*If a mass of fluid be at rest, any portion of it may be supposed to become solid without affecting its equilibrium or the pressure of the surrounding fluid.*

For there will be no alteration in the forces acting on the fluid, and the action between the solidified portion and the rest of the fluid, or between the solidified portion and any surface with which it may be in contact, will still be normal to its surface (Art. 4); therefore the equilibrium of the solid can be considered as maintained by the external forces which act upon it, and the pressure of the remaining fluid.

This proposition enables us to employ the principles of statics in the discussion of the equilibrium of fluids.

6. Measure of the Pressure of Fluids.—The pressure of a fluid on a plane is measured, when uniform over the plane, by the force exerted on a unit of area. Consider a mass of fluid at rest under the action of any forces, and let A be the area of a plane surface in contact with the fluid, and P the force which is required to counterbalance the action of the fluid upon A. Then if the action of the fluid upon A be uniform, $\dfrac{P}{A}$ is the pressure on each unit of the area A, and this is usually represented by p.

If the pressure be variable, as, for instance, on the vertical side of a vessel, it must be considered as varying continuously from point to point of the area A, and the pressure at any point is measured by that which would be exerted on a unit of area, supposing the pressure over the whole unit to be exerted at the same rate as at the point considered. If we suppose the area A, and the pressure P, to diminish indefinitely, the pressure may be regarded as uniform on the infinitesimal area dA, and we shall have $\dfrac{dP}{dA} = p$ to express the *rate* of pressure at the point considered.

By the rate of pressure at a *point* is meant the force which would be exerted on a unit of area, if the rate of pressure over the unit were uniform and the same as at the point considered.

7. The Pressure at any Point of a Fluid at Rest is the same in every Direction.

—By this statement is meant that, if at any point of a fluid, there be placed a small plane area containing the point, the pressure of the fluid upon the plane at that point will be independent of the position of the plane.

This is the most important of the characteristic properties of a fluid. It is often established by experiments; it may, however, be deduced, independently of experiments, in the following manner:

Let a small tetrahedron of fluid be supposed solidified (Art. 5); then it is kept at rest by the pressures on its faces, which are always normal (Art. 4), and by the impressed * forces on its mass. The pressures on the faces depend on the *areas* of the faces, and the impressed forces depend on the *volume* and density. When the fluid is considered homogeneous, the former forces vary as the *square*, and the latter vary as the *cube* of one of the edges of the solid; supposing therefore the solid to be indefinitely diminished, while it always retains a similar form, the latter forces, being small quantities of the third order, vanish in comparison with the pressures on the faces, which are small quantities of the second order; and hence these pressures form a system of forces in equilibrium.

Let p, p_1 be the rates of pressure (Art. 6) on the faces, ABD, BCD, and resolve these forces parallel and perpendicular to the edge AC: let β and γ be

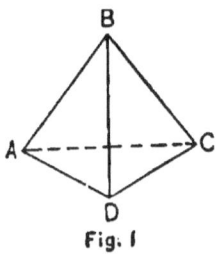

Fig. 1

---

* See Anal. Mechs., Art. 234.

the angles which a plane perpendicular to AC makes with the planes, ABD and BCD, respectively; then we have

$$p \cdot ABD \cdot \cos \beta = p_1 \cdot BCD \cdot \cos \gamma. \qquad (1)$$

But $ABD \cdot \cos \beta = BCD \cdot \cos \gamma =$ the projections of the areas ABD and BCD on a plane perpendicular to AC; therefore (1) becomes

$$p = p_1.$$

And similarly it may be shown that the pressures on the other two faces are each equal to p or p_1. As the tetrahedron may be taken with its faces in any direction, it follows that the pressure at any point is the same in every direction.*

COR.—Hence the lateral pressure of a fluid at any point is equal to its perpendicular pressure.

SCH.—This property constitutes a remarkable distinction between fluids and solids, the latter pressing with their whole weight in the direction of gravity alone. This property of fluids can be conceived to arise only from the extreme facility with which the particles move among one another. It is not easy to imagine how this can take place, if the particles be supposed to be in immediate contact; they are therefore probably kept at a distance from one another by some repulsive force.

8. Equal Transmission of Fluid Pressure.—(1) Let AB be a tube of uniform bore, and of any shape whatever, filled with a liquid, and closed at its extremities by two pistons A and B, which fit the bore exactly, but yet can move along it with perfect freedom; and let the interior of the tube be perfectly smooth, so as not to offer the least resistance to

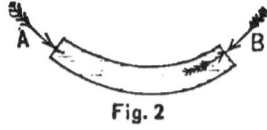

Fig. 2

---

* See Besant's Hydromechanics, p. 4.

the motion of the liquid along it. Then it may be assumed as self-evident, that if any force be applied to the piston A, perpendicular to its surface, and directed inwards, it will push the liquid forward, and thus produce a pressure on the piston B, which will drive it out of the tube, unless there be an equal force at B, pushing in the opposite direction, to counteract the force at A and keep the liquid at rest. This property of liquids is a direct result of experiment.

(2) Let ABCD represent a closed vessel of any shape, filled with a liquid ; let A and B be any two points in the surface of the vessel, and let two circular holes be made at these points, having the same area; into these let two short tubes be inserted, each tube entering a little way into the liquid, and provided with a piston that fits it accurately, and which may move within it with the utmost freedom. Now suppose that the two orifices, A and B, are connect-

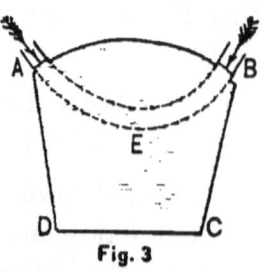

Fig. 3

ed by a tube of liquid AEB, in the interior of the vessel, of uniform bore, and of any form, and imagine all the liquid in the vessel, except that contained in the tube, to be solidified. This will not affect the equilibrium (Art. 5). But, under these circumstances, if a pressure be applied to the piston A, and directed inwards, it will, as shown in (1) above, be transmitted to B, and will require an equal force at B to counteract it and keep the fluid at rest.

If we suppose the piston B, to be taken *anywhere* on the surface, it is evident from what has been said that any pressure applied to the piston A will be transmitted to B, and will require an equal pressure at B to counteract it. It is also evident that if we have *several* openings, each equal to B, closed by pistons, any pressure applied to one piston will be transmitted undivided to every other piston, and will require an equal pressure at each of those pistons to counteract it. The above reasoning remains true, no matter where we suppose the point B to be taken. Hence *any pressure, applied to the surface of an incompressible fluid at rest, is transmitted equally to all parts of the fluid and to its whole surface.*

COR.—If a point E, be within a liquid, the pressure transmitted from the piston A, to a plane surface of given area, and having its centre at E, is constant for every possible position of the plane, and is always perpendicular to it.

9. The Pressures on Two Pistons are in Equilibrium when Proportional to their Areas. — Let Fig. 4 represent a vessel with two apertures, in which pistons are fitted; and let the vessel be filled with any liquid. Now, any pressure applied to the small piston p, will be transmitted by the liquid to the large piston P, so that every portion of surface in the large piston will be pressed upwards with the same force that an equal portion of surface in the small piston is pressed downwards (Art. 8). Let a = the area of the piston p, A = the area of the piston P, p = the whole pressure applied to the small piston p, and P = the whole pressure produced upon the large piston P; then, since the whole pressure on the large piston is equal to that on the small one taken as many times as the area of the small one is contained in that of the large, we have for equilibrium,

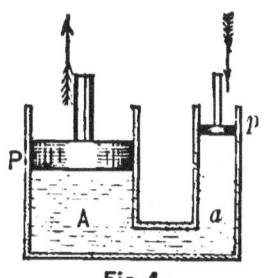

Fig. 4

$$P = p \times \frac{A}{a}; \quad \text{or,} \quad \frac{P}{p} = \frac{A}{a}. \tag{1}$$

That is, *two forces applied to pistons which are connected with each other through the intervention of some confined liquid, will be in equilibrium when* they are *directly proportional to the areas of the pistons upon which they act.*

This result is wholly independent of the relative dimensions and positions of the pistons. Let a be the unit of area, say a square inch or square foot, then will p be the pressure applied to the unit of area, and (1) becomes

$$P = pA. \qquad (2)$$

That is, *the pressure transmitted to any portion of the surface of the vessel is equal to that applied to the unit of surface multiplied by the area of the surface to which the pressure is transmitted.*

If the area of the piston P be one square foot, and a pressure of 10 lbs. is applied at the piston p, it follows from (2) that a pressure of 1440 lbs. will be transmitted to the piston P, and this must be counteracted by a pressure of 1440 lbs. on that piston. Also, the interior of the vessel will sustain an outward pressure of 10 lbs. on every square inch of its surface. And if the pressure on the piston p, is increased till the vessel bursts, the fracture is as likely to occur in some other part as in that towards which the force is directed.

Cor.—If in the vessel (Fig. 4) the piston A, be made sufficiently large, the pressure transmitted from a to A may be increased indefinitely; a very great weight upon A may be raised by a small pressure at a, the weight lifted being greater in proportion to the size of A, or inversely to the size of a. To increase the upward force at A, we must enlarge the surface of A or diminish the surface of a, and the only limitation to the increase of the force at A will be the want of sufficient strength in the vessel to resist the increased pressure.

On this principle, machines of immense mechanical power are constructed, which will be described in a future chapter.

EXAMPLES.

1. If the area of the piston a be a square inch, and if it be pressed by a force of 25 lbs., find the pressure which will be transmitted to a surface of 35 square inches.
Ans. 875 lbs.

2. If the area of the piston be 3 square inches, and if the pressure on it be 96 lbs., find the pressure which will be transmitted to a surface of 17.5 square inches.
Ans. 560 lbs.

3. If the area of the piston be 2.5 sq. in., and if the pressure on it be 50 lbs., what pressure will this transmit to a portion of the surface of the vessel whose shape is circular and whose diameter is one foot? *Ans.* 2261.95 lbs.

10. Pressure of a Liquid at any Depth.—Thus far only the transmission of *external* pressures has been considered; we shall now determine the effects of the *internal* pressure due to the weight of the particles of the liquid itself.

Let DAE be the surface of the liquid at rest, and take any point B, in the liquid; draw BA vertically to the surface, and describe a small cylinder about BA with its base horizontal. Imagine this cylinder to become solid (Art. 5). Then this solid body is at rest under its own weight, the pressure of the fluid on the end B, and the fluid pressures on the curved surface.

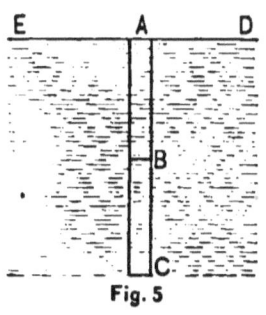
Fig. 5

The fluid pressures on the curved surface are all horizontal (Art. 4), and the fluid pressure on the end B, and the weight of the solid are vertical forces, and each group is separately in equilibrium. Hence the fluid pressure on B must be equal to the weight of the solid AB; if a be the area of the base $AB = z$, w the weight of a unit of volume, and p the pressure at B, we have

$$pa = waz; \quad \text{or,} \quad p = wz; \quad (1)$$

that is, *the pressure at any depth varies as the depth below the surface.*

Similarly, let B and C be any two points in the same vertical line, and let the cylinder BC, be solidified; then, from what has just been shown, the pressures at B and C must differ by the weight of the cylinder BC, *i. e.*, the pressure at C is greater than that at B by the weight of a column of liquid whose base is equal to the area C, and whose height is BC.

Hence, if p and p' be the pressures at B and C, and $BC = z$, we have

$$p'a - pa = waz; \quad \text{or,} \quad p' - p = wz; \quad (2)$$

that is, *the difference of the pressures at any two points varies as the vertical distance between the points.*

Cor. 1.—If W be the weight of a mass M, of fluid, then (Anal. Mechs., Art. 24), we have

$$W = Mg. \quad (3)$$

If V be the volume of the mass M, of fluid, and ρ be its density, then (Anal. Mechs., Art. 11), we have

$$M = V\rho. \quad (4)$$

$$\therefore W = g\rho V. \quad (5)$$

For a unit of volume we have $V = 1$, therefore (5) becomes

$$W = g\rho.$$

From (1) we have,

$$pa = waz = W = g\rho V \text{ [from (5)]}.$$

or,
$$pa = g\rho az \text{ (since } V = az\text{)}; \quad (6)$$

$$\therefore p = g\rho z. \quad (7)$$

COR. 2.—If A be the area of the base of a vessel, h its height, and P the whole pressure on the base, we have, from (6),

$$P = g\rho hA. \quad (8)$$

That is, *the pressure of a liquid on any horizontal area is equal to the weight of a column of the liquid whose base is equal to the area, and whose height is equal to the height of the surface of the liquid above the area.*

It is evidently immaterial whether the surface pressed is that of the base of the vessel or a horizontal surface of an immersed solid.

COR. 3.—Since the weight of a cubic foot of water = 1000 ozs. = 62.5 lbs., we have, for the pressure on the bottom of any vessel containing water,

$$P = 62.5hA \text{ lbs.,} \quad (9)$$

where h is the height in feet of the surface of the water above the base, and A the area of the base in square feet.

COR. 4.—*The pressure on the base of any vessel is independent of the form of the vessel.*

Thus, if a hollow cone, vertex upwards, be filled with water, and if r be the radius of the base and h the height of the cone, we have for the pressure on the base,

$$P = g\rho \pi r^2 h \text{ [from (8)],}$$

or,
$$P = 62.5\pi r^2 h \text{ [from (9)] ;}$$

that is, the pressure on the base is the same as if the cone were a cylinder of liquid of the same base and height as the cone; the pressure is three times the weight of the enclosed water.

This increased pressure on the base is caused by the reaction of the curved surface of the cone. The pressure on the curved surface consists of an assemblage of forces whose vertical components all point downwards and react upon the base.

EXAMPLES.

1. If a surface of one square inch be placed in a vessel completely filled with water, and if the pressure upon it be 2 lbs., what will be the pressure on one square inch placed at a level 75 inches lower?

Here A = one square inch, h = 75 inches, and P and P' are the pressures at the upper and lower points; therefore we have, from (2) and (8),

$$P' - P = 252.5^* \times 75$$
$$= 18937.5 \text{ grains}$$
$$= 2.705 \text{ lbs.}$$

$$\therefore P' = 2.705 + 2 = 4.705 \text{ lbs.}$$

2. If the pressure on the upper surface, whose area is a circle of half an inch radius, is 1.5 lbs., find the pressure on another circular area whose radius is one inch, placed at a depth 10 feet lower in the water. *Ans.* 19.5986 lbs.

11. The Free Surface of a Liquid at Rest is a Horizontal Plane.—Let ABCD represent the section of a vessel containing a liquid subject to the action of gravity; then will its free surface be horizontal. For, if the free surface is not horizontal, suppose it to be the curved line, APB. Take any point P, of the surface where the tangent to the curve is not horizontal; let

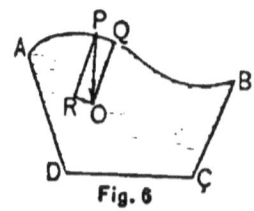

Fig. 6

---

* The weight of one cubic inch of water at the standard temperature is 252.5 grains.

the vertical line PO, be drawn to represent the weight of the particle of liquid at P, and resolve this weight into two components PR and PQ, the former perpendicular, and the latter parallel to the surface. The first of these is opposed by the reaction of the surface; the second, being unopposed, causes the particle to move downwards to a lower level. It is evident, therefore, that if the free surface be one of equilibrium, it must at each point be perpendicular to the direction of gravity, *i. e.*, it must be horizontal.

COR. 1.—Since the directions of gravity, acting on particles remote from each other, are convergent to the earth's centre, nearly, large surfaces of liquids are not plane, but curved, and conform to the general figure of the earth. But, for small areas of surface the curvature cannot be detected, because the deviation from a plane is infinitesimal.

COR. 2.—The pressure of the atmosphere is found to be about 14.73 lbs. to a square inch, or very nearly 15 lbs. The pressure, therefore, on any given area can be calculated, and if π be the atmospheric pressure on the unit of area, the pressure at a depth z of a liquid, the surface of which is exposed to the pressure of the atmosphere, will be, from (7) of Art. 10,

$$p = g\rho z + \pi. \tag{1}$$

COR. 3.—Since the pressures are equal when the depths are equal (Art. 10), it follows that the areas of equal pressure are also areas of equal depth; therefore, since the surface of a liquid is a horizontal plane, an area of equal pressure is everywhere at the same depth below a horizontal plane, *i. e.*, *an area of equal pressure is a horizontal plane; and, conversely, the pressure of a liquid at rest at all points of a horizontal plane is the same.*

Hence it appears that when the pressure on the surface of a liquid is either zero or is equal to the constant atmospheric pressure, all points on its surface must be in the same horizontal plane, even though the continuity of the surface be interrupted by the immersion of solid bodies. *If any number of vessels, containing the same liquid, are in communication, the liquid stands at the same height in each vessel.*

This sometimes appears under the form of the assertion that *liquids maintain their level.*

REM.—The construction by which towns are supplied with water furnishes a practical illustration of this principle. Pipes, leading from a reservoir placed on a height, carry the water, underground or over roads, to the tops of houses or to any point provided that no portion of a pipe is higher than the surface of the water in the reservoir.

12. The Common Surface of Two Fluids.—Let AD
be the upper surface of the lighter fluid, and BC the common surface of the two fluids; AD is horizontal (Art. 11). Let P and Q be two points in the heavier liquid, both equally distant from the surface AD, and therefore in the same horizontal plane. Draw the vertical lines Pa and Qb, meeting the common surface of the fluids in c and d. Let w be the weight of a unit of volume of the upper fluid, and w' that of the lower.

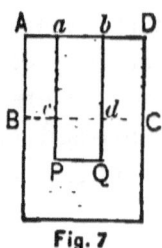
Fig. 7

Then we have

$$\text{pressure at } P = w' \cdot cP + w \cdot ac;$$

and $$\text{pressure at } Q = w' \cdot dQ + w \cdot bd.$$

Since the pressures at P and Q are equal (Art. 11, Cor. 3), they being in the same horizontal plane, we have

$$w' \cdot cP + w \cdot ac = w' \cdot dQ + w \cdot bd. \qquad (1)$$

But $$cP + ac = dQ + bd \qquad (2)$$

multiplying (2) by w, and subtracting the result from (1), we have

$$(w' - w)\, c\mathrm{P} = (w' - w)\, d\mathrm{Q},$$

$$\therefore\ c\mathrm{P} = d\mathrm{Q},$$

and hence BC is horizontal.

That is, *the common surface of two fluids that do not mix is a horizontal plane.*

Cor.—This proposition is true, whatever be the number of fluids; the common surfaces are all horizontal. If, therefore, the number be infinite, or the density of the fluid vary according to any law, the surface of each will still be horizontal.*

13. Two Fluids in a Bent Tube.—Let A and C be the two surfaces, B the common surface, and ρ, ρ' the densities of AB and BC. Let z and z' represent the heights of the surfaces A and C, above the common surface B, and take B′ in the denser fluid in the same horizontal plane as B.

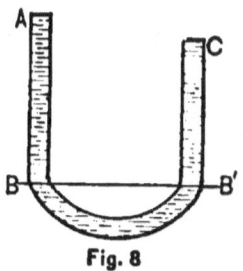

Fig. 8

Then we have,

the pressure at B $= g\rho z$ [(7) of Art. 10];

the pressure at B′ $= g\rho' z'$,

and these are equal (Art. 11, Cor. 3).

$$\therefore\ g\rho z = g\rho' z',$$

$$\therefore\ z : z' :: \rho' : \rho.$$

Hence, *when two fluids that do not mix together meet in a bent tube, the heights of their upper sur-*

---

* See Besant's Hydrostatics, p. 31; also Bland's Hydrostatics, p. 20.

faces above their common surface are inversely proportional to their densities.*

14. Pressure on Planes.—*To find the pressure on a plane area in the form of a rectangle when it is just immersed in a liquid, with one edge in the surface, and its plane inclined at an angle θ to the vertical.*

Let ABCD be a vertical section perpendicular to the plane of the rectangle; then AB is the section of the surface of the liquid, and AC ($= a$) is the section of the rectangle, the upper edge b, of the rectangle being in the surface of the liquid perpendicular to AC at A.

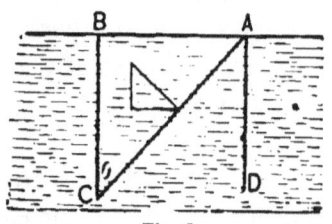

Fig. 9

Pass a vertical plane BC, through the lower edge of the rectangle, and suppose the fluid in ABC to become solid. The weight of this solid is supported by the plane AC, since the pressure on BC is horizontal (Art. 4). Let R be the normal pressure on the plane AC; resolving R horizontally and vertically, we have, for vertical forces,

$$R \sin \theta = \text{weight of ABC}$$
$$= g\rho \cdot \tfrac{1}{2} AB \cdot BC \cdot b \quad [(5) \text{ of Art. 10}]$$
$$= \tfrac{1}{2} g\rho a^2 b \sin \theta \cos \theta.$$

$$\therefore \ R = g\rho ab \cdot \tfrac{1}{2}a \cos \theta; \qquad (1)$$

that is, *the pressure on the rectangle is equal to the weight of a column of fluid whose base is the rec-*

---

* The common barometer may be considered as an example of this principle. The air and mercury are the two fluids. If the atmosphere had the same density throughout as at the surface of the earth, its height could be determined. For height of mercury in barometer : height of air :: density of air : density of mercury. As mercury is 10784 times as dense as air, the height of the atmosphere would be 10784 × 30 inches, or nearly 5 miles.

tangle, and whose height is equal to the depth of the middle point of the rectangle below the surface.

Cor.—When $\theta = 0$, (1) becomes

$$R = g\rho ab \cdot \tfrac{1}{2}a \qquad (2)$$
$$= g\rho \,(\text{area BC}) \,(\text{depth of middle of BC}),$$

which *is the pressure on the vertical plane* BC; hence the law is the same as for the inclined plane AC.

15. The Whole Pressure.—*The whole pressure of a fluid on any surface with which it is in contact is the sum of the normal pressures on each of its elements.*

If the surface is a *plane*, the pressure at every point is in the same direction, and the whole pressure is the same as the resultant pressure. If it is a *curved* surface, the whole pressure is the arithmetic sum of all the pressures acting in various directions over the surface. The following proposition is general, and applies to curved or plane surfaces, for unit-area.

Let S be the surface, and p the pressure at a point of an element dS, of the surface. Then

$$pdS = \text{the pressure on the element}; \qquad (1)$$

and since the pressure is the same in every direction (Art. 7), p will be the normal pressure on this element, whatever be its position or inclination. Hence,

$$\iint p dS = \text{the whole normal pressure}, \qquad (2)$$

the integration extending over the whole of the surface considered.

If gravity be the only force acting on the fluid,* we have, from (7) of Art. 10,

$$p = g\rho z, \qquad (3)$$

z being measured vertically and positive downwards from the surface of the liquid. From (2) and (3) we have,

$$\iint p\,dS = \iint g\rho z\,dS. \qquad (4)$$

Calling $\bar{z}$ the depth of the centre of gravity of the surface S, below the surface of the liquid, we have [Anal. Mechs., Art. 84, (1), ρ and k being constant],

$$\bar{z}\cdot S = \iint z\,dS,$$

which in (4) gives,

$$\iint p\,dS = g\rho\bar{z}S, \qquad (5)$$

for the whole pressure on the surface S. That is, *the whole pressure of a liquid on any surface is equal to the weight of a cylindrical column of the liquid whose base is a plane area equal to the area of the surface and whose height is equal to the depth of the centre of gravity of the surface below the surface of the liquid.*

REM.—The student will now be able to appreciate more clearly the nature of fluid pressures, and to see that the action of a fluid does not depend upon its *quantity*, but upon the *position* and *arrangement* of its *continuous portions*. It must be borne in mind that the surface of an incompressible fluid or liquid is always the horizontal plane drawn through the highest point or points of the fluid, and that the pressure on any area depends only on its depth below that horizontal plane (Art. 10). For instance, in the construction of dock-gates, or canal-locks, it is not the

---

* The fluid being a homogeneous liquid.

expanse of sea outside which will affect the pressure, but the *height* of the surface of the sea.

EXAMPLES.

1. If a cubical vessel be filled with a liquid, find the ratio of the pressures against the bottom and one of its sides.

The area of the surface pressed, in each case, is the same, but the depth of the centre of gravity of the bottom is twice that of the centre of gravity of the side; therefore the ratio is 2 : 1.

2. Find the pressure on the internal surface of a sphere when filled with water.

Let $a =$ the radius of the sphere; then the area of the surface $= 4\pi a^2$, and the depth of the centre of gravity of the surface below the surface of the water $= a$; therefore, calling the pressure P, we have, from (5),

$$P = g\rho a \cdot 4\pi a^2 = 4g\rho\pi a^3,$$

which is three times the weight of the water.

3. A rectangle is immersed with two opposite sides horizontal, the upper one at a depth c, and its plane inclined at an angle θ to the horizontal. Find the whole pressure on the plane.

[Let a be the horizontal side, and b the other side.]

$$Ans. \text{ Pressure} = g\rho ab\left(c + \frac{b}{2}\sin\theta\right).$$

4. If a cubical vessel is filled with water, and each edge of the vessel is 10 ft., find the pressure on the bottom and on a side, a cubic foot of water weighing 62½ lbs.

$$Ans. \begin{cases} \text{Pressure on bottom} = 62500 \text{ lbs.} \\ \text{Pressure on side} = 31250 \text{ lbs.} \end{cases}$$

5. A rectangular surface, 10 ft. by 5 ft., is immersed in water with its short sides horizontal, the upper side being

20 ft. and the lower 26 ft. below the surface of the water. Find the pressure it sustains. *Ans.* 32 tons.*

6. A cylinder, closed at both ends, is immersed in a liquid so that its axis is inclined at an angle θ, to the horizon, and the highest point of the cylinder just touches the surface of the liquid. Find the whole pressure on the cylinder, including its plane ends.

[Let $r =$ the radius of the base and $h =$ the length of the cylinder.] *Ans.* $g\rho\pi r (h + r)(h \sin \theta + 2r \cos \theta)$.

7. A hemispherical cup is filled with water, and placed with its base vertical. Find the pressures on the curved and plane surfaces.

Ans. $\begin{cases} \text{Pressure on the curved surface} = 2g\rho\pi a^3. \\ \text{Pressure on the plane surface} = g\rho\pi a^3. \end{cases}$

This example shows the distinction between the total pressure of a fluid on a curved surface, and on that portion of it which is perpendicular to any given plane. The pressure on the vertical plane side of the hemispherical cup might be obtained by finding the sum of the horizontal components of the actual pressures on all the elements of the curved surface. This latter pressure, called the *resultant horizontal pressure* of the liquid on the surface, is equal to the pressure of the liquid on the plane base, otherwise the cup would have a tendency to move in a horizontal direction.

16. Centre of Pressure.—*The centre of pressure of a plane area immersed in a fluid is the point of action of the resultant fluid pressure upon the plane area.* It is therefore that point in an immersed plane surface or side of a vessel containing a fluid, to which, if a force equal and opposite to the resultant of all the press-

---
* One ton = 2240 lbs.

ures upon it be applied, this force would keep the surface at rest.

In the case of a liquid, it is clear that the centre of pressure of a horizontal area, the pressure on every point of which is the same, is its centre of gravity; and since the pressure varies as the depth (Art. 10), the centre of pressure of any plane area, not horizontal, is below its centre of gravity.

Let ABCD be any immersed plane area; take the rectangular axes OX and OY, in the plane of the area. Let (x, y) be any point P, of the area referred to these axes, and p the pressure at this point, and let EH be the line of intersection of the plane with the surface of the fluid.

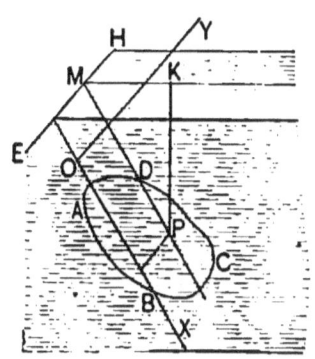

Fig. 10

Then the pressure on the element of area

$$= p \, dx \, dy;$$

∴ the resultant pressure $= \int\int p \, dx \, dy.$

Let $(\bar{x}, \bar{y})$ be the centre of pressure; then the moment of the resultant pressure about OY

$$= \bar{x} \int\int p \, dx \, dy;$$

and the sum of the moments of the pressures on all the elements of area about OY

$$= \int\int px \, dx \, dy.$$

Therefore, since the moment of the resultant pressure is equal to the sum of the moments of the component pressures (**Anal. Mechs.**, Art. 59), we have

$$\bar{x}\int\int p\,dx\,dy = \int\int px\,dx\,dy;$$

$$\therefore\ \bar{x} = \frac{\int\int px\,dx\,dy}{\int\int p\,dx\,dy}, \tag{1}$$

and, similarly,
$$\bar{y} = \frac{\int\int py\,dx\,dy}{\int\int p\,dx\,dy}, \tag{2}$$

the integration extending over the whole of the area considered.

If polar co-ordinates be used, a similar process will give the equations,

$$\bar{x} = \frac{\int\int pr^2\cos\theta\,dr\,d\theta}{\int\int pr\,dr\,d\theta}, \tag{3}$$

$$\bar{y} = \frac{\int\int pr^2\sin\theta\,dr\,d\theta}{\int\int pr\,dr\,d\theta}. \tag{4}$$

If the fluid be homogeneous and incompressible, and if gravity be the only force acting on it, we have [Art. 10, (7)].

$$p = g\rho h,$$

where h ($=$ PK) is the depth of the point P below the surface of the fluid, K being the projection of P on this surface, and KM being perpendicular to EH. Substituting this value of p in (1) and (2), we get

$$\bar{x} = \frac{\int\int hx\,dx\,dy}{\int\int h\,dx\,dy}, \tag{5}$$

$$\bar{y} = \frac{\int\int hy\,dx\,dy}{\int\int h\,dx\,dy}, \qquad (6)$$

If we take for the axis of y the line of intersection EH, of the plane with the surface of the fluid, and denote the inclination of the plane to the horizon by θ, we have

$$PK = PM \sin PMK,$$

or, $\qquad h = x \sin \theta;$

which in (5) and (6) give us,

$$\bar{x} = \frac{\int\int x^2\,dx\,dy}{\int\int x\,dx\,dy}, \qquad (7)$$

$$\bar{y} = \frac{\int\int xy\,dx\,dy}{\int\int x\,dx\,dy}. \qquad (8)$$

COR. 1.—If the axis of x be taken so that it will be symmetrical with respect to the immersed plane, the pressures on opposite sides of this axis will obviously be equal, and the centre of pressure will be on this axis, or $\bar{y} = 0$.

COR. 2.—Since (7) and (8) are independent of θ it appears that the centre of pressure is independent of the inclination of the plane to the horizon, so that if a plane area be immersed in a fluid, and then turned about its line of intersection with the surface of the fluid as a fixed axis, the centre of pressure will remain unchanged.

REM.—The position of the centre of pressure is of great importance in practical problems. It is often necessary to know the exact effect of the pressure exerted by fluids against the sides of vessels and obstacles exposed to their action, in order to adjust the dimensions of the latter, so that they may be strong enough to resist this pressure. Examples are furnished us in the construction of reservoirs, in which large quantities of water are collected and retained for purposes of irrigation, the supply of cities and towns, or to drive machinery, and of dykes to protect low districts from being inundated by seas and lakes and rivers in times of freshets.

EXAMPLES.

1. Find the centre of pressure of a rectangle vertically immersed, and having one side parallel to the surface of the fluid, and at a given distance below it.

Let a and b be the distances of the bottom and top of the rectangle from the surface of the fluid, and d the width; take the intersection of the plane of the rectangle with the surface of the fluid for the axis of y, and the middle point of this side for the origin, the axis of x bisecting the rectangle. Then from (7) we have,

$$\bar{x} = \frac{\int_b^a \int_0^{\frac{1}{2}d} x^2 \, dx \, dy}{\int_b^a \int_0^{\frac{1}{2}d} x \, dx \, dy} = \frac{\int_b^a x^2 \, dx}{\int_b^a x \, dx}$$

$$= \frac{2}{3} \frac{a^3 - b^3}{a^2 - b^2}.$$

Cor.—If the upper side of the rectangle is in the surface of the fluid, $b = 0$, and therefore we have

$$\bar{x} = \tfrac{2}{3}a,$$

or the centre of pressure of a vertical rectangle, one side being in the surface of the fluid, is two-thirds the height of the rectangle below the surface of the fluid. The value of $\bar{y}$ is evidently zero.

2. Find the centre of pressure of an isosceles triangle whose base is horizontal and opposite vertex in the surface of the fluid.

Let a be the altitude of the triangle and b its base. Take the intersection of the plane of the triangle with the surface of the fluid for the axis of y and the vertex for the origin, the axis of x bisecting the triangle. Then from (7) we have,

$$\bar{x} = \frac{\int_0^a \int_0^{\frac{b}{2a}x} x^2 \, dx \, dy}{\int_0^a \int_0^{\frac{b}{2a}x} x \, dx \, dy} = \frac{\int_0^a x^3 \, dx}{\int_0^a x^2 \, dx} = \tfrac{3}{4}a.$$

3. A quadrant of a circle is just immersed vertically in a fluid, with one edge in the surface. Find its centre of pressure.

Take the edge in the surface for the axis of y, and the vertical edge for the axis of x, and let a be the radius. Then, from (7) and (8), we have

$$\bar{x} = \frac{\int_0^a \int_0^{\sqrt{a^2-x^2}} x^2 \, dx \, dy}{\int_0^a \int_0^{\sqrt{a^2-x^2}} x \, dx \, dy} = \frac{\int_0^a x^2 (a^2 - x^2)^{\frac{1}{2}} \, dx}{\int_0^a x (a^2 - x^2)^{\frac{1}{2}} \, dx}$$

$$= \frac{a^4 \pi}{16} \div \frac{a^3}{3} = \frac{3}{16} a \pi \, ;$$

and $\quad \bar{y} = \dfrac{\int_0^a \int_0^{\sqrt{a^2-x^2}} xy \, dx \, dy}{\int_0^a \int_0^{\sqrt{a^2-x^2}} x \, dx \, dy} = \dfrac{\tfrac{1}{2} \int_0^a x(a^2 - x^2) \, dx}{\int_0^a x(a^2 - x^2)^{\frac{1}{2}} \, dx}$

$$= \frac{a^4}{8} \div \frac{a^3}{3} = \frac{3}{8} a.$$

(See Besant's Hydromechanics, p. 41.)

4. Find the centre of pressure of the triangle in Ex. 2, when it is inverted so that the base is in the surface of the fluid.

Ans. At a distance of $\frac{1}{4}a$ below the surface of the fluid.

5. An immersed rectangle has two sides horizontal, the inclination of the plane of the rectangle to the horizon is θ, the depth of its upper side below the surface of the fluid is c, the sides of the rectangle are a and b, the latter horizontal. Find its centre of pressure.

[Take the upper side for the axis of y, and its middle point for the origin.]

Ans. $\bar{x} = \dfrac{a}{3} \cdot \dfrac{3c + 2a \sin \theta}{2c + a \sin \theta}$ and $\bar{y} = 0$.

17. Embankments.—An embankment generally consists of a large mass of earth and other material. When used for the side of a reservoir or canal, to bank up a river,* to keep out the sea,† or in general to dam back water, they are constructed on certain principles, and are opposed to the effort made by the water to spread itself. The effort to overthrow the embankment arises from the force which the water exerts horizontally; and the stability is caused by the weight of the embankment. When therefore there is an equilibrium, the former of these forces must be equivalent to the latter.

An embankment is generally made wider than is absolutely necessary, to give strength and stability sufficient to insure it against all risks. Frequently they slant only on the side that is away from the water. In every case the embankment should be built much stronger at the bottom than at the top, for the pressure of water increases as the depth.

18. Embankment when the Face on the Water Side is Vertical.—Find the stability of an embankment

---

* Called dykes. † Called sea-walls.

whose section has the form of a trapezoid when the water stands at its brim.

Let ABCD be the cross-section of the embankment; draw DE parallel to the vertical side BC; let G and g be the centres of gravity of the rectangle and triangle respectively; draw the vertical lines GH and gK; let AB $= a$, DC $= b$, BC $= h$, $w =$ the weight of each cubic foot of the material, and $w_1 =$ the weight of a cubic foot of water.

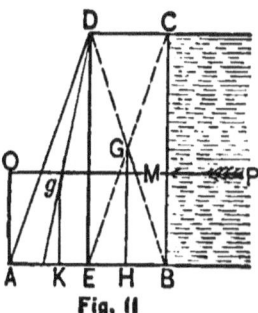
Fig. 11

The forces acting are the weight of the wall, and the fluid pressure on BC. As the embankment is uniform throughout its length, and also the pressure on it, we may determine the stability by taking only one foot in length. Take BM $= \frac{1}{3}$BC, and M will be the centre of pressure (Art. 16, Ex. 1, Cor.). The resultant P, of the pressure of the water against the wall acts at M, and tends to turn the embankment over its outer edge A. Hence, we have

the moment of P = pressure of water on BC × AO
(Art. 15)
$$= \tfrac{1}{2}h^2 w_1 \times \tfrac{1}{3}h = \tfrac{1}{6}h^3 w_1 ; \qquad (1)$$

the moment of AED = weight of AED × AK
$$= \tfrac{1}{2}(a-b)hw \times \tfrac{2}{3}(a-b)$$
$$= \tfrac{1}{3}(a-b)^2 hw ; \qquad (2)$$

the moment of EBCD = weight of EBCD × AH
$$= bhw \times (a-\tfrac{1}{2}b) ; \qquad (3)$$

∴ the moment of ABCD $= [\tfrac{1}{3}(a-b)^2 + b(a-\tfrac{1}{2}b)] hw.$ (4)

If the embankment be upon the point of overturning on A, the moments in (1) and (4) are equal to each other, and we have

$$\tfrac{1}{3}h^3 w_1 = [\tfrac{1}{3}(a-b)^2 + b(a-\tfrac{1}{2}b)]\,hw,$$

or,
$$h^2 = [2(a-b)^2 + 3b(2a-b)]\frac{w}{w_1}, \qquad (5)$$

and the embankment will be overturned or not, according as

$$h > \text{ or } < \sqrt{[2(a-b)^2 + 3b(2a-b)]\frac{w}{w_1}}.$$

COR.—If the embankment is rectangular, $b = a$, and (5) becomes

$$h^2 = 3a^2 \frac{w}{w_1}. \qquad (6)$$

If the embankment is triangular, $b = 0$, and (5) becomes

$$h^2 = 2a^2 \frac{w}{w_1}.$$

19. Embankment when the Face on the Water Side is Slanting.

—Find the stability of an embankment whose section is a trapezoid which slants on both sides, viz., towards the water and away from it.

(1) Suppose the embankment to yield to the pressure of the fluid by *turning round the outer edge A.*

Let ABCD be the cross-section of the embankment. Since the pressure of a fluid is always normal to the surface with which it is in contact (Art. 4), the pressure on the slanting face BC, of this embankment is inclined to the horizon, and hence the stability of the embankment is caused by its weight and the vertical pressure of the fluid on the face BC, while the effort to overthrow it is caused by the horizontal pressure of the fluid.

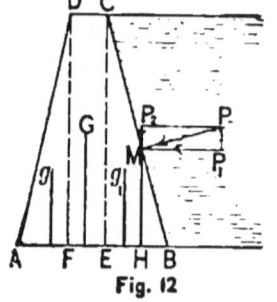

Fig. 12

Let P_1 and P_2 be the horizontal and vertical components of the normal pressure P, and α the angle which the direction of the normal pressure makes with the horizon; then we have, for the horizontal component,

$$P_1 = P \cos \alpha$$
$$= \text{area of } BC \times \tfrac{1}{2}CE \times w_1 \cos \alpha \text{ (Art. 15)}$$
$$= \text{area of } CE \times \tfrac{1}{2}hw_1,$$

where $h = CE$, and w_1 is the weight of a cubic foot of the fluid.

Similarly, $P_2 = \text{area of } BE \times \tfrac{1}{2}hw_1$;

but area of CE is the projection of BC on CE, and area of BE is the projection of CB on EB; *i. e., the pressure exerted by a fluid in any direction upon a surface is equal to the weight of a column of the fluid, whose base is the projection of the surface at right angles to the given direction, and whose height is the depth of the centre of gravity of the surface below the surface of the fluid.*

Hence, since the projection at right angles to the vertical direction is the horizontal projection, and that at right angles to a horizontal direction is a vertical one, we find the *vertical pressure* of the fluid against a surface by treating its *horizontal projection* as the surface pressed upon, and, on the contrary, the *horizontal pressure* of the fluid in any direction by treating the *vertical projection* of the surface at right angles to the given direction as the surface pressed upon, and in both cases we must regard the depth of the centre of gravity of the surface below the surface of the fluid as the "height of the column."

Let g, G, and g_1 be the centres of gravity of AFD, FECD, and EBC; let $AB = a$, $DC = b$, $AF = c$, $EB = d$, and $w =$ the weight of each cubic foot of the embankment. The horizontal pressure of the water acting at M tends to turn the embankment over its outer edge A. Hence, we have

the moment of $P_1 = \tfrac{1}{2}h^2 w_1 \times \tfrac{1}{3}h = \tfrac{1}{6}h^3 w_1$; (1)

the moment of $P_2 = d \times \tfrac{1}{2}hw_1 \times AH$
$= \tfrac{1}{2}hdw_1\,(a - \tfrac{1}{2}d)$;

the moment of AFD $=$ wt. of ADF $\times \tfrac{2}{3}AF$
$= \tfrac{1}{2}chw \times \tfrac{2}{3}c = \tfrac{1}{3}c^2 hw$;

the moment of FECD $=$ wt. of FECD $\times (AF+\tfrac{1}{2}FE)$
$= bhw \times (c+\tfrac{1}{2}b)$;

the moment of EBC $=$ wt. of EBC $\times (AB - \tfrac{2}{3}BE)$
$= \tfrac{1}{2}dhw \times (a - \tfrac{2}{3}d)$
$= \tfrac{1}{6}dhw\,(3a - 2d)$.

$\therefore$ the moment of ABCD $= \left[\dfrac{c^2}{3} + \dfrac{1}{2}b(2c+b)\right.$
$\left.+ \dfrac{d}{6}(3a - 2d)\right]hw + \dfrac{d}{6}(3a - d)\,hw_1$. (2)

If the embankment be upon the point of overturning on A, the moments in (1) and (2) are equal to each other, and we have

$\tfrac{1}{6}h^3 w_1 = \left[\dfrac{c^2}{3} + \dfrac{b}{2}(2c+b) + \dfrac{d}{6}(3a-2d)\right]hw + \dfrac{d}{6}(3a-d)\,hw_1$

or, $h^2 = [2c^2 + 3b(2c+b) + d(3a-2d)]\dfrac{w}{w_1} + d(3a-d)$, (3)

and the embankment will be overturned or not, according as

$h >$ or $< \sqrt{[2c^2 + 3b(2c+b) + d(3a-2d)]\dfrac{w}{w_1} + d(3a-d)}$.

Cor. 1.—If the embankment is of the form of Fig. 11, $d = 0$, and (3) becomes

$$h^2 = [2c^2 + 3b(2c+b)]\frac{w}{w_1}, \qquad (4)$$

which agrees with (5) of Art. 18.

Cor. 2.—If the embankment is rectangular, $c = 0$, and (4) becomes

$$h^2 = 3b^2\frac{w}{w_1},$$

which agrees with (6) of Art. 18.

(2) Suppose the embankment to yield to the pressure of the fluid by *sliding* along the horizontal base AB.

The horizontal pressure of the fluid, from (1), is

$$P_1 = \tfrac{1}{2}h^2 w_1;$$

the vertical pressure of the fluid is

$$P_2 = \tfrac{1}{2}dhw_1.$$

The weight of the embankment is

$$\frac{a+b}{2}hw;$$

and the entire vertical pressure of the embankment and the water on its face is

$$\frac{a+b}{2}hw + \tfrac{1}{2}dhw_1$$
$$= (aw + bw + dw_1)\tfrac{1}{2}h.$$

Let $\mu =$ the coefficient of friction; then the friction between the embankment and the surface of the ground on which it rests is (Anal. Mechs., Art. 92),

$$(aw + bw + dw_1)\tfrac{1}{2}h\mu.$$

When the horizontal pressure of the water pushes the embankment forward, we must have

$$\tfrac{1}{2}h^2 w_1 = (aw + bw + dw_1)\tfrac{1}{2}h\mu;$$

or, more simply, $h = \left[(a + b) \dfrac{w}{w_1} + d \right] \mu,$ (5)

and the dam will move or not according as

$$h > \text{ or } < \left[(a + b) \dfrac{w}{w_1} + d \right] \mu.$$

Cor.—If the embankment is rectangular, $d = 0$ and $b = a$, and (5) becomes

$$h = 2a \dfrac{w}{w_1} \mu.$$

20. Pressure upon Both Sides of a Surface.—If a plane surface is subjected on *both sides* to the pressure of a fluid, the two resultants of the pressures on the two sides have a new resultant, which, as they act in opposite directions, is obtained by subtracting one from the other.

Let AB be a flood-gate with the water pressing on both sides of it, to determine the resultant pressure, and the centre of pressure. Let $AB = a$, the depth of the water on one side; $DB = b$, the depth of the water on the other side; $P =$ the resulting pressure on the gate; and $w_1 =$ the weight of a cubic foot of water. Then

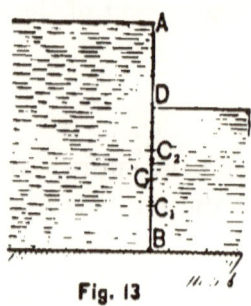

Fig. 13

$P =$ pressure on AB $-$ pressure on DB ;

$\therefore \ P = \tfrac{1}{2} (a^2 - b^2) w_1.$ (1)

Now let C and C_1 be the centres of pressure of the surfaces AB and DB, and C_2 the point to which the resultant pressure P, is applied. Then, taking moments with respect to A, and putting $AC_2 = \bar{z}$, we have

$P \times \bar{z} =$ pressure on AB $\times$ AC $-$ pressure on DB $\times$ AC$_1$
$= \tfrac{1}{2} a^2 w_1 \times \tfrac{2}{3} a - \tfrac{1}{2} b^2 w_1 (a - \tfrac{2}{3} b).$

$$\therefore \bar{z} = \frac{2a^3 + b^3 - 3ab^2}{3(a^2 - b^2)} \quad \text{[from (1)]}$$

$$= \frac{2a^2 + 2ab - b^2}{3(a+b)}. \tag{2}$$

EXAMPLES.

1. The total breadth of a flood-gate is $2b$ feet, and the depth is a feet; the hinges are placed at d feet from the respective extremities of the gate; required the pressure upon the lower hinge.

Let AB represent the height of the gate, D and E the hinges, and C the centre of pressure of the water. The pressure of the water upon each half of the gate $= \frac{1}{2}a^2bw$; and since the pressure of the water at C is supported by the hinges D and E, we have, by the equality of moments with respect to D,

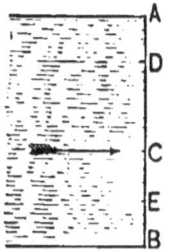

Fig. 14

Pressure on $E \times DE$ = Pressure on $C \times DC$;

but $DE = a - 2d$, and $DC = \frac{2}{3}a - d$;

$\therefore$ Pressure on $E (a - 2d) = \frac{1}{2}a^2bw (\frac{2}{3}a - d)$;

$\therefore$ Pressure on $E = \dfrac{a^2bw(2a - 3d)}{6(a - 2d)}$.

2. A brick wall, with rectangular cross-section, 12 ft high and 3 ft. thick, sustains the pressure of water against one of its faces. Find the height to which the water may rise without overthrowing the wall, each cubic foot of the wall weighing 112 lbs.

Ans. 8.34 ft., or within 3.66 ft. of the top of the wall.

3. A brick wall, whose cross-section is a right-angled triangle, weighs 120 lbs. per cubic foot, and sustains the pressure of water against its vertical face; its height is

14 ft., and its base is 6 feet. Show that the wall will be overthrown by the pressure of water against it, when it rises to the top of the wall.

21. Rotating Liquid.—It has been shown (Art. 11) that, if a liquid at rest be subject to the force of gravity only, its free surface must be horizontal, *i. e.*, everywhere perpendicular to the direction of gravity. In the same way it may be shown that, if a liquid be subject to any forces whatever, its surface, if free, must at every point be perpendicular to the resultant of the forces which act upon that point. For, if the resultant had any other direction, it could be resolved into two components, one in the direction of the normal and the other in the direction of the tangent; the first of these would be opposed by the reaction of the surface; the second, being unopposed, would cause the particle to move, which is contrary to the hypothesis that the surface is at rest: hence the surface is at every point perpendicular to the resultant of the forces which act upon that point.

If a quantity of liquid in a vessel be made to rotate uniformly about a vertical axis, the surface of the liquid will take the form of a paraboloid of revolution.

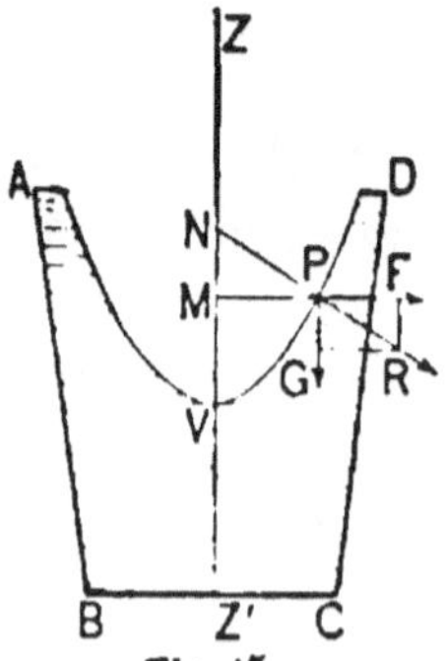

Let ABCD represent a vertical section made by a plane passing through ZZ', the axis of rotation of the vessel containing the liquid, and let the curved line AVD, represent the section of the surface of liquid made by this plane, and let P be any point taken on this section.

Now every particle of the liquid moves uniformly in a horizontal circle whose centre is in the axis ZZ', and therefore is urged horizontally by a centrifugal force directed

Fig. 15

from the axis. Let m be the mass of the particle at P, ω the angular velocity of the liquid, and γ the distance MP, and denote the centrifugal force by P; then (Anal. Mechs., Art. 198) we have, for the centrifugal force on the particle m,

$$P = m\omega^2\gamma. \qquad (1)$$

The particle is also urged vertically downwards by its own weight mg, due to the force of gravity; hence the particle is in equilibrium under the action of gravity mg, of the centrifugal force $m\omega^2\gamma$, and of the reaction of the surface of the liquid which is normal, and therefore the resultant of mg and $m\omega^2\gamma$ must be normal to the surface.

Let PF and PG represent the centrifugal force and force of gravity, respectively; then, completing the parallelogram of forces, the resultant of these PR, must be normal to the surface at P. Let this normal meet the axis in N; since the triangles, GPR and MNP, are similar, we have

$$\text{NM : MP :: PG : GR } (= \text{PF});$$

or $\quad$ NM : γ :: mg : $m\omega^2\gamma$;

$$\therefore \text{NM} = \frac{g}{\omega^2}. \qquad (2)$$

But NM is the subnormal of the curve, AVD; therefore

the subnormal NM $= \frac{g}{\omega^2} = $ a constant,

which is a property of the parabola. Hence the curve AVD, is a parabola whose latus rectum is $\frac{2g}{\omega^2}$, and therefore *the surface is a paraboloid of revolution.*

SCH.—It will be seen that this result is independent of the form of the containing vessel. The axis of rotation, in fact, may be within or without the fluid, but in any case it will be the axis of the surface of the paraboloid.

EXAMPLES.

1. If the vessel (Fig. 15) contain a liquid, and make 30 revolutions per minute, find the value of NM.

Here $\omega = 2\pi \times 30 \div 60 = \pi$, and $g = 32$; therefore we have, from (2),

$$NM = \frac{32}{\pi^2} = 3.242 \text{ ft.} = 38.9 \text{ in.}$$

2. If the vessel make one turn in a second, find the value of NM. *Ans.* 9.72 in.

3. If the vessel make 95 turns per minute, find the value of NM. *Ans.* 3.88 in.

22. The Pressure at any Point of a Rotating Liquid.—Let ABCD be a vertical section through the axis of a vessel containing a rotating liquid; let Q be any point (x, y) in the liquid referred to the rectangular axes OX, OY, and describe a small vertical prism having Q in its base, which is to be taken horizontal.

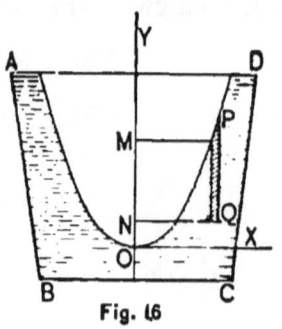

Fig. 16

The prism PQ of liquid rotates uniformly under the action of the pressure around it, but its weight is entirely supported by the vertical pressure on its base. Hence, if p be the pressure, and ρ the density, we have

$$p = g\rho \text{PQ}. \qquad (1)$$

But [Art. 21, (2)],

$$PQ = OM - ON = \frac{\omega^2 \overline{QN}^2}{2g} - ON,$$

which in (1) gives,

$$p = \rho\left(\tfrac{1}{2}\omega^2 x^2 - gy\right), \qquad (2)$$

which gives the pressure at any point in terms of the angular velocity and of the co-ordinates of the point referred to the axis and vertex of the paraboloid. (See Besant's Hydrostatics, p. 152.)

Cor.—If Q be lower than O, y is negative, and (2) becomes

$$p = \rho (\tfrac{1}{2}\omega^2 x^2 + gy). \tag{3}$$

EXAMPLES.

1. A tube ABCD, the equal branches of which are vertical, and BC horizontal, is filled with liquid and made to rotate uniformly about the axis of AB; find how much liquid will flow out of the end D.

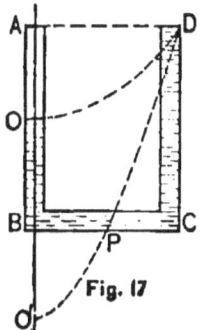

Fig. 17

The liquid will flow out until the surface in AB is the vertex of a parabola passing through D, and having its axis vertical and latus rectum = $\frac{2g}{\omega^2}$ (Art. 21).

If then O be the vertex of the parabola, we shall have

$$\overline{AD}^2 = \frac{2g}{\omega^2} AO;$$

$$\therefore \quad AO = \frac{\omega^2}{2g} \overline{AD}^2,$$

which gives AO, and thus determines the quantity which flows out.

If, however, AO be greater than AB, *i. e.*, if O be *below* B, at O', for instance, the surface of the liquid will be in BC, at P. We shall then have,

$$\overline{AD}^2 = \frac{2g}{\omega^2} AO';$$

and
$$\overline{BP}^2 = \frac{2g}{\omega^2} BO';$$

$$= \overline{AD}^2 - \frac{2g}{\omega^2} AB,$$

which determines the position of P. (Besant's Hydrostatics, p. 154.)

2. A straight tube AB, filled with liquid, is made to rotate about a vertical axis through A; find how much flows out at B.

Ans. All above P, where P is tangent to the parabola whose latus rectum is $\frac{2g}{\omega^2}$ and whose axis is coincident with the vertical line through A, and $AP = \frac{g}{\omega^2} \cot \alpha \csc \alpha$, where α is the angle OAB.

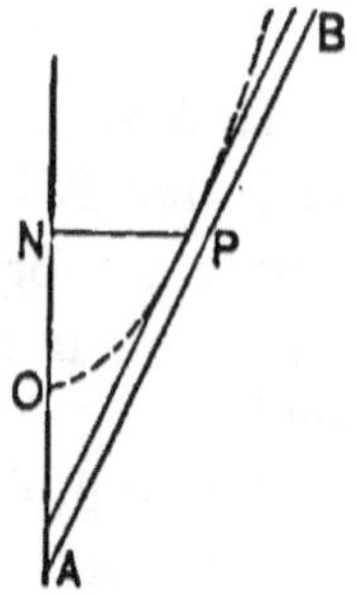

Fig. 18

23. Strength of Pipes and Boilers.

An important application of the theory of the pressure of fluids is the determination of the thickness of *pipes, boilers,* etc. In order that these vessels shall be strong enough to resist the pressure of the liquid, their *walls* must be made of a certain thickness, which depends upon the pressure of the liquid and the internal diameter of the vessel.

Let it be required to find the thickness of a pipe of any material necessary to resist a given pressure.

A cylindrical vessel may burst either transversely or longitudinally; but the former is less likely to occur than the latter, as appears from the following investigation.

(1) *When the rupture is transverse.*

Let ABCD (Fig. 19) be a section of pipe perpendicular to its axis, the interior surface of which is subjected to a pressure of p on each unit of surface. Let $2r$ be the diameter MD of the interior, then will the surface pressed be measured by πr^2, which is the area of the cross-section of the interior, and the whole pressure upon the surface of the end of the pipe and which produces rupture will be measured by

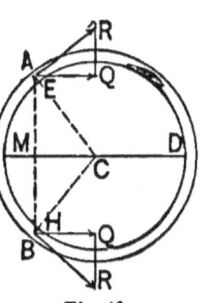

Fig. 19

$$\pi r^2 p. \qquad (1)$$

Let $e = AE =$ the thickness of the pipe; then the cross-section of the material of the pipe

$$= \pi (r + e)^2 - \pi r^2 = \pi e (e + 2r).$$

Let T denote the strength of the material of which the pipe is composed, for each unit of cross-section; then the strength of the entire pipe in the direction of the axis

$$= \pi e (e + 2r) T, \qquad (2)$$

and since the whole pressure in (1) when rupture is about to take place must be held in equilibrium by the strength in (2), we have

$$\pi e (e + 2r) T = \pi r^2 p,$$

$$\therefore e = \frac{rp}{2\left(1 + \frac{e}{2r}\right)T} = \frac{rp}{2T}, \qquad (3)$$

since e is usually very small in comparison with $2r$.

(2) *When the rupture is longitudinal.*

Let EMH be any portion of the wall whose length is l, and let $2\alpha =$ the angle ECH. Then, since the projection of EMH at right angles to the line MD passing through the centre is a rectangle whose area $= 2rl \sin \alpha$, the mean pressure of the fluid on the wall, EMH

$$= 2rl \sin \alpha \, p \text{ (Art. 19)}. \qquad (4)$$

Now this pressure must be held in equilibrium by the forces of cohesion, R, R, acting tangentially on the cross-sections, AE and BH, of the wall of the pipe. Denoting the components of R, R, parallel to MD, by Q, Q, we have

$$2Q = 2R \sin \alpha = 2elT \sin \alpha, \qquad (5)$$

e being the thickness of the pipe and T the strength of each unit of section.

Therefore, from (4) and (5) we have,

$$2elT \sin \alpha = 2rlp \sin \alpha;$$

$$\therefore \; e = \frac{rp}{T}, \qquad (6)$$

which shows that the thickness of the pipe is independent of its length.

Otherwise thus, by the principle of work.

The whole surface of the interior of the pipe $= 2\pi rl$; and the whole pressure upon the surface $= 2\pi rlp$. Suppose the pipe to rupture longitudinally,* under this pressure, its radius becoming $r + dr$; then the path described by the pressure will be dr, and the work done by the pressure

$$= 2\pi rlp \, dr. \qquad (7)$$

The force R, which resists rupture and acts tangentially, $= eTl$. While the radius of the interior changes from r to $r + dr$, the circumference changes from $2\pi r$ to $2\pi(r+dr)$; then the path described by the resistance $= 2\pi \, dr$, and the work done by the resistance

$$= 2\pi eTl \, dr. \qquad (8)$$

---

* Longitudinal tension produces transverse rupture, and transverse tension produces longitudinal rupture. The stretching tendency to rupture longitudinally is a transverse stretching, *i. e.*, the pipe tends to bulge out all along its length; hence, transversely, r becomes $r + dr$.

Therefore, from (7) and (8), by the principle of work, we have
$$2\pi e Tl\, dr = 2\pi r l p\, dr,$$
$$\therefore\ e = \frac{rp}{T},$$
which is the same as (6).

From (3) and (6) it follows that, *to prevent a longitudinal rupture, the wall must be made twice as thick as would be necessary to prevent a transverse one.*

Cor.—Since $p = zw$ [from (1) of Art. 10], (3) and (6) become, respectively,
$$e = \frac{rp}{2T} = \frac{rzw}{2T}; \quad \text{and} \quad e = \frac{rp}{T} = \frac{rzw}{T};$$
that is, *the thickness of similar pipes must vary directly as their diameter and as the pressure upon the unit of surface, or in the case of a liquid, as the depth of the pipe below the upper surface of the liquid, and inversely as the strength of each unit of section.*

A pipe which has twice the diameter, and has to sustain four times the pressure of another, must be eight times as thick. (See Weisbach's Mechs., Vol. I., p. 739; Bartlett's Mechs., p. 294; Tate's Mechs., p. 268.)

EXAMPLES.

1. It is found that the pressure is uniform over a square yard of a plane area in contact with fluid, and that the pressure on the area is 13608 lbs.; find the measure of the pressure at any point (Art. 6), (1) when the unit of length is an inch, (2) when it is two inches.

Ans. (1) $10\tfrac{1}{2}$ lbs.; (2) 42 lbs.

2. If the area of a (Fig. 4) be a square inch, and if it be pressed by a force of 15 lbs., what pressure * will this transmit to the piston A if its diameter be 10 in.?

Ans. Pressure on A $= 1178$ lbs.

3. If the diameter of a be 4 in., and if the pressure on it be 185 lbs., what pressure will be exerted on A if its area is one square foot? *Ans.* Pressure on A $= 2120$ lbs.

4. If the area of a be 20 square inches, and if it be pressed by a force of 360 lbs., find the diameter of A so that it shall be pressed upwards by a force of 10 tons (one ton $= 2240$ lbs.) *Ans.* Diameter of A $= 39.8$ in.

5. If the diameter of A (Fig. 3) be one inch, and if the surface at E be a square whose side is one-quarter of an inch, find the pressure transmitted to E if that on A be 10 lbs. *Ans.* Pressure on E $= 0.795$ lbs.

6. If the area of A be $2\frac{1}{2}$ sq. in., and the pressure on it 56 lbs., find the pressure transmitted to a surface at E, the area of which is a triangle whose base is $\frac{3}{8}$ of an inch, and whose height is $\frac{1}{10}$ of an inch.

Ans. Pressure on E $= 0.42$ lbs.

7. A cylindrical pipe which is filled with water opens into another pipe the diameter of which is three times its own diameter; if a force of 20 lbs. be applied to the water in the smaller pipe, find the force on the open end of the larger pipe which is necessary to keep the water at rest.

Ans. 180 lbs.

8. Required (1) the pressure on the sides of a cubical vessel filled with water, and (2) the pressure on the bottom, the side of the vessel being a ft. (Art. 10).

Ans. (1) $125a^3$ lbs.; (2) $62.5a^3$ lbs.

9. A cylindrical vessel is filled with water; the height of the vessel is a ft., and the diameter of the base d feet. Find (1) the pressure upon the side and (2) the pressure on the bottom. *Ans.* (1) $31\frac{1}{4}\pi a^2 d$; (2) $15\frac{5}{8}\pi a d^2$.

---

* In the first seven examples, the weight of the liquid itself is not considered.

10. Find the height of the vessel in Ex. 9 so that the pressure on the side may be equal to the pressure on the bottom.

Ans. The height must equal the radius of the base.

11. The pressure on a square inch of surface in a vessel of mercury is 1000 grains. Find the pressure on a circular surface of one-quarter inch radius, placed 9 in. lower down, mercury being 13.5 times as heavy as water.

Ans. Pressure = 0.8886 lbs.

12. The water in a canal lock rises to a height of 18 ft. against a gate whose breadth is 11 ft. Find the total pressure against the gate. *Ans.* Pressure = $49\frac{1}{4}$ tons.*

13. The upper side of a sluice-gate is $10\frac{1}{2}$ ft. beneath the surface; its dimensions are 3 ft. vertical by 18 in. horizontal. Find the pressure upon it.

Ans. Pressure = $1\frac{1}{2}$ tons.*

14. A dyke to shut out the sea is 200 yards long, and is built in courses of masonry one foot high; the water rises against it to a height of 6 fathoms. Find the pressure against the 1st, 18th, and 36th courses.

Ans. $\begin{cases} \text{1st pressure} = 610.4 \text{ tons.*} \\ \text{2d pressure} = 318.1 \text{ tons.} \\ \text{3d pressure} = 8.6 \text{ tons.} \end{cases}$

15. Find the pressure, in pounds, of a cylinder of water 4 inches in diameter and 45 ft. in height.

Ans. Pressure = 244.8 lbs.

16. A cubical vessel, each side of which is 10 ft., is filled with water, and a tube 32 ft. long is fitted to an aperture in it, whose area is one square inch. If the tube be vertical, and of the same size as the aperture, and filled with water, find the pressure on the interior surface of the vessel, (1) neglecting the weight of the water it contains, (2) when the weight of the water is taken into account.

Ans. (1) 1,200,000 lbs.; (2) 1,387,500 lbs.

---

* One ton = 2240 lbs.

17. Find the pressure on a square inch at a depth of 100 ft. in a lake, (1) neglecting, (2) taking account of the atmospheric* pressure. *Ans.* (1) $43\frac{49}{72}$ lbs.; (2) 58 lbs.

18. A reservoir of water is 200 ft. above the level of the ground floor of a house; find the pressure of the water, per square inch, in a pipe at a height of 30 ft. above the ground floor, neglecting atmospheric pressure. *Ans.* $73\frac{11}{72}$ lbs.

19. An equilateral triangular area is immersed vertically in water with a side, one foot in length, in the surface. Find the pressure upon it in ounces. *Ans.* 125 oz.

20. A hollow cone, vertex upwards, is just filled with liquid. Find (1) the pressure on its base, (2) the normal pressure on its curved surface, (3) the vertical pressure on the curved surface. [Let $r =$ the radius of the base and $h =$ the altitude.]

Ans. (1) $g\rho\pi r^2 h$; (2) $\frac{2}{3}g\rho\pi r h\sqrt{r^2+h^2}$; (3) $\frac{2}{3}g\rho\pi r^2 h$.

21. A vertical rectangle has one side in the surface of a liquid. Divide it by a horizontal line into two parts on which the pressures are equal.

Ans. If h be the vertical side, the depth of the horizontal line $= \dfrac{h}{\sqrt{2}}$.

22. A vertical triangle, altitude h, has its base horizontal and its vertex in the surface. Divide it by a horizontal line into two parts on which the pressures are equal.

Ans. The depth $= \dfrac{h}{\sqrt[3]{2}}$.

23. A smooth vertical cylinder one foot in height and one foot in diameter is filled with water, and closed by a heavy piston weighing 4 lbs. Find the whole pressure on its curved surface. *Ans.* $16 + \dfrac{125\pi}{4}$ lbs.

---

* See Art. 11, Cor. 2.

24. A hollow cylinder, closed at both ends, is just filled with water and held with its axis horizontal; if the whole pressure on its surface, including the plane ends, be three times the weight of the fluid, compare the height and diameter of the cylinder. *Ans.* As $1:1$.

25. The side AB of a triangle ABC is in the surface of a fluid, and a point D is taken in AC, such that the pressures on the triangles BAD, BDC, are equal. Find the ratio AD : DC. *Ans.* As $1 : \sqrt{2} - 1$.

26. The diameters of the two pistons, p and P (Fig. 4), are $2\tfrac{1}{4}$ in. and 9 in., respectively, and the smaller is 60 in. above the larger. What force must be applied to the smaller piston that the larger may exert a pressure of 1600 lbs. ? *Ans.* 112.8 lbs.

27. Compare the pressure on the area of a parabola with that on its circumscribing rectangle, both being immersed perpendicularly to the vertex. *Ans.* As $4:5$.

28. A cubical vessel is filled with two liquids, of given densities, the volume of each being the same. Find the pressure on the base and on any side of the vessel.

Let a be a side of the vessel, ρ and ρ' the densities of the upper and lower liquids, ρ' being greater than ρ.

The pressure on the base = the weight of the whole fluid
$$= g \frac{a^3}{2} (\rho' + \rho).$$

The pressure on the upper half of any side
$$= g\rho \frac{a^2}{2} \cdot \frac{a}{4} = \frac{1}{8} g\rho a^3.$$

To find the pressure on the lower half, replace the upper liquid by an equal weight of the lower liquid, which will not affect the pressure at any point of the lower half. If a' be the height of this equal weight, we have

$$\rho'a' = \rho\frac{a}{2};$$

and the depth of the centre of gravity of the lower half below the upper surface of the equal weight

$$= a' + \frac{a}{4} = \frac{a}{4}\left(1 + \frac{2\rho}{\rho'}\right);$$

therefore, the pressure on the lower half

$$= g\rho'\frac{a^2}{2}\cdot\frac{a}{4}\left(1 + \frac{2\rho}{\rho'}\right)$$

$$= \tfrac{1}{8}ga^3(\rho' + 2\rho).$$

(Besant's Hydrostatics, p. 37.)

29. A circle is just immersed vertically in a fluid. Find on which chord, drawn from the lowest point, the pressure is the greatest.

[Let ADBC be the circle with radius a and BC the required chord, which bisect in H, and draw HK perpendicular to AB; ∴ etc.]   *Ans.* AK $= \tfrac{1}{3}a$.

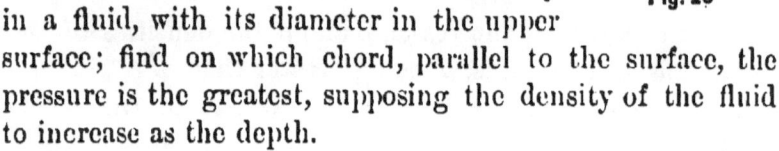

Fig. 20

30. A semicircle is immersed vertically in a fluid, with its diameter in the upper surface; find on which chord, parallel to the surface, the pressure is the greatest, supposing the density of the fluid to increase as the depth.

[Let LBM (Fig. 20) be the semicircle, and DE the chord on which the pressure is the greatest, and a the radius of the circle. Then if the density were uniform, the pressure would vary as DG × GF (Art. 15); but, since the density varies as the depth, the pressure varies as DG × GF²; ∴ etc.]   *Ans.* FG $= a\sqrt{\tfrac{2}{3}}$.

31. If LBM (Fig. 20) be a parabola, FB $= b$, the latus rectum $= 4a$, and the other conditions the same as in Ex. 30, find FG, the depth of the chord of greatest pressure below the upper surface.   *Ans.* FG $= \tfrac{4}{5}b$.

32. The lighter of two fluids, whose densities are as 2 : 3, rests on the heavier, to a depth of 4 in. A square is immersed in a vertical position, with one side in the upper surface. Determine the side of the square in order that the pressures on the portions in the two fluids may be equal.

Ans. $\frac{4}{3}(1 + \sqrt{10})$ in.

33. Find the centre of pressure of a semi-parabola, the extreme ordinate coinciding with the surface of the fluid.

[Let LBF (Fig. 20) be the semi-parabola; let BF $= a$, and LF $= b$, and suppose O to be the centre of pressure, OG being parallel to LF.] *Ans.* FG $= \frac{4}{7}a$; GO $= \frac{6}{16}b$.

34. A quadrant of a circle is just immersed vertically in a liquid, with one edge in the surface, as in Ex. 3, Art. 16. Find the centre of pressure when the density varies as the depth.

Taking the edge in the surface for the axis of y and the vertical edge for the axis of x, we find

$$\bar{x} = \frac{32}{15}\frac{a}{\pi}; \qquad \bar{y} = \frac{16}{15}\frac{a}{\pi}.$$

35. The total breadth of a water passage closed by a pair of flood-gates is 10 ft. and its depth is 6 ft. ; the hinges are placed at one foot from the top and bottom. Find the pressure upon the lower hinge when the water rises to the top of the gates. *Ans.* $4218\frac{3}{4}$ lbs.

36. If we suppose everything to be the same as in Ex. 2, Art. 20, except that the height of the wall is determined by the condition that the wall just sustain the pressure when the water rises to the top, what is the height of the wall?

Ans. 6.96 ft.

37. A wall of masonry, a section of which is a rectangle, is 10 ft. high, 3 ft. thick, and each cubic foot weighs 100 lbs. Find the greatest height of water it will sustain without being overturned. *Ans.* $6\sqrt[3]{2}$.

38. If the height of the wall be 8 ft., its thickness 6 ft., and each cubic foot weighs 180 lbs., find whether it will stand or fall when the water is on a level with the top.
Ans.

39. The depth AB of the water in the head bay (Fig. 13) is 7 ft., the depth DB of the water in the chamber of the lock is 4 ft., and the width of the lock-chamber is 7.5 ft. ; find (1) the resultant pressure upon the gate AB, and (2) the depth of the point of application of the resultant pressure below the surface of the water in the head bay.
Ans. (1) 7734.4 lbs.; (2) 4.18 ft.

40. If the vessel (Fig. 15) make 140 turns per minute, find the value of NM. *Ans.* 1.78 in.

41. A hollow paraboloid of revolution, with its axis vertical and vertex downwards, is half filled with liquid. With what angular velocity must it be made to rotate about its axis, in order that the liquid may just rise to the rim of the vessel.
Ans. If $2p =$ latus rectum, $\omega^2 = \dfrac{g}{2p}$.

42. If the vessel in the last example be filled with liquid, find the angular velocity and the time of rotation that it may just be emptied.
Ans. If $2p =$ latus rectum, $\omega^2 = \dfrac{g}{p}$; time $= 2\pi\sqrt{\dfrac{p}{g}}$.

43. A hemispherical bowl is filled with liquid, which is made to rotate uniformly about the vertical radius of the bowl. Find how much runs over.
Ans. $\dfrac{1}{4}\dfrac{\pi a^4 \omega^2}{g}$.

44. A closed cylindrical vessel, height h and radius a, is just filled with liquid, and rotates uniformly about its vertical axis. Find the pressures on its upper and lower ends, and the whole pressure on its curved surface.
Ans. $g\rho\pi a^2 \dfrac{a^2\omega^2}{4g}$, $g\rho\pi a^2 \left(\dfrac{a^2\omega^2}{4g} + h\right)$, and $g\rho\pi ah\left(h + \dfrac{\omega^2 a^2}{g}\right)$.

CHAPTER II.

EQUILIBRIUM OF FLOATING BODIES. — SPECIFIC GRAVITY.

24. Upward Pressure, Buoyant Effort.—*To find the resultant pressure of a liquid on the surface of a solid either wholly or partially immersed.*

Let ABCD be a solid floating in a liquid whose upper surface is EF. Imagine this solid removed, and the space it occupied filled with the liquid, and suppose this liquid to be solidified. It is clear that the resultant pressure upon this solidified liquid will be the same as upon the original solid. But this solidified mass is at rest under the action of its own weight and the pressure of the surrounding liquid; and, as its own weight acts vertically downward through its centre of gravity, the resultant pressure of the surrounding liquid must be equal to the weight of the solidified mass, and must act vertically upwards in a line passing through its centre of gravity.

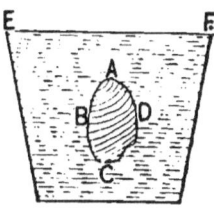

Fig. 21

The above reasoning is equally applicable to the case of a body immersed in elastic fluid.

Therefore, *if a solid be either wholly or partially immersed in a fluid, it loses as much of its weight as is equal to the weight of the fluid it displaces.**

---

* The discovery of this principle is due to Archimedes. (Goodeve, p. 190; Galbraith, p. 49.)

COR. 1.—If a body be supported entirely by a fluid, the weight of the body must be equal to the weight of the fluid displaced, and the centres of gravity of the body and of the fluid displaced must lie in the same vertical line.

SCH.—These conditions hold good, whatever be the nature of the fluid in which the body is floating. If it be heterogeneous, the displaced fluid must consist of horizontal strata of the same kind as, and continuous with, the horizontal strata of uniform density, in which the particles of the surrounding fluid are necessarily arranged. If, for instance, a solid body float in water, partially immersed, its weight will be equal to the weight of the water displaced, together with the weight of the air displaced.

The upward pressure of a fluid against a solid, and which is equal to the weight of the displaced fluid, is called the *buoyant effort* of a fluid. The centre of gravity of the displaced fluid is called *the centre of buoyancy*. The buoyant effort exerted by a fluid acts vertically upwards through the centre of buoyancy.

The enunciation and proof of this proposition are due to Archimedes, and it is a remarkable fact in the history of science, that no further progress was made in Hydrostatics for 1800 years, and until the time of Stevinus, Galileo, and Torricelli, the clear idea of fluid action thus expounded by Archimedes remained barren of results.

An anecdote is told of Archimedes, which practically illustrates the accuracy of his conceptions. Hiero, king of Syracuse, had a certain quantity of gold made into a crown, and suspecting that the goldsmith had abstracted some of the gold and used a portion of alloy of the same weight in its place, he applied to Archimedes to investigate whether such was the case, and to ascertain the nature of the alloy. It is related that while Archimedes was in his bath, reflecting over the difficult problem which the king had given him, he observed the water running over the sides of the bath, and it occurred to him that he was displacing a quantity of water equal in volume to that of his own body, and therefore that a quantity of pure gold equal in weight to the crown would displace less water than the crown, the volume of any weight of alloy being greater than that of an equal weight of gold.

He concluded at once that he could completely solve the king's problem, by weighing the crown in water. Overjoyed with his discovery, he ran directly into the street, crying out, "Eureka! Eureka!"

The two books of Archimedes which have come down to us were first found in old Latin MS. by Nicholas Tartaglia, and edited by him in 1537. These books contain the solutions of a number of problems on the equilibrium of paraboloids, and various problems relating to the equilibrium of portions of spherical bodies.

The authenticity of these books is confirmed by the fact that they are referred to by Strabo, who not only mentions their title, but also quotes from the first book.

25. Conditions of Equilibrium of an Immersed Solid.—Let v denote the volume and ρ the density of the solid; v' the volume and ρ' the density of the displaced fluid: the weights of the solid and of the displaced fluid will be respectively $g\rho v$ and $g\rho'v'$; then, if the solid rest in equilibrium in the fluid, we shall have

$$g\rho v = g\rho'v'. \tag{1}$$

If we suppose the solid to be entirely immersed, the volumes v and v' will be equal, and the densities ρ and ρ' must also be equal if the solid remains in equilibrium, having no tendency either to ascend or descend.

But if the weight of the immersed solid be greater than that of the fluid displaced, we shall have

$$g\rho v > g\rho'v;$$

and the solid will be urged *downwards* by a force equal to $g\rho v - g\rho'v$.

If, on the contrary, the weight of the solid be less than that of the fluid, we shall have

$$g\rho v < g\rho'v;$$

and the solid will be urged *upwards* by a force equal to $g\rho'v - g\rho v$.

That is, *the wholly immersed solid will descend, remain at rest, or ascend, according as its density is greater than, equal to, or less than the density of the fluid.*

In the first case the solid will descend to the bottom, and press it with a force equal to the excess of its weight above that of an equal bulk of fluid.

In the third case the solid will rise to the surface, and be but partially immersed, the volume v' of the fluid displaced by the solid having the same weight as the entire solid.

[An egg, placed in a vessel of fresh water, sinks to the bottom of the vessel, its mean density being a little greater than that of the water. If, instead of fresh water, salt water is employed, the egg floats at the surface of the liquid, which is a little denser than the egg. If fresh water is carefully poured on the salt water, a mixture of the two liquids takes place where they are in contact; and if the egg is put in the upper part, it will descend, and, after a few oscillations, remain at rest in a layer of liquid of which it displaces a volume whose weight is equal to its own.]

Cor.—From (1) we have

$$v : v' :: \rho' : \rho;$$

therefore, *if a homogeneous solid float in a fluid, its whole volume is to the volume of the displaced fluid as the density of the fluid is to the density of the solid.*

Sch.—When the floating solid and fluid are both homogeneous, the centre of gravity of the part immersed will coincide with the centre of buoyancy.

The section of a floating body formed by the plane of the surface of the fluid in which the body floats is called *the plane of flotation*. The line passing through the centre of gravity of the floating body and the centre of buoyancy is called the *axis of flotation*.

The weight $g\rho v$ of the body acting downwards, and the buoyant effort $g\rho' v'$ acting upwards (Art. 24, Sch.), form a couple, by which the body rotates till the directions of these forces coincide, *i.e.*, till the centre of gravity of the body and the centre of buoyancy come into the same vertical line.

EXAMPLE.

1. A piece of oak containing 32 cubic inches, floats in water; how much water will it displace, the density of the oak being 0.743 times that of water?

$\qquad\qquad\qquad\qquad\qquad$ *Ans.* 23.776 cu. in.

26. Depth of Flotation.*—*The depth to which a body sinks below its plane of flotation is called its Depth of Flotation.* When the form and weight of a floating body are known, its *depth of flotation* can be calculated.

Denoting the volume and density of the body by v and ρ, and of the displaced fluid by v' and ρ', respectively, we have [Art. 25, (1)].

$$g\rho v = g\rho' v';$$

$$\therefore\ v' = \frac{\rho}{\rho'} v, \qquad (1)$$

by which the depth of flotation can be determined, whenever v' can be determined in terms of that depth.

EXAMPLES.

1. Let the solid be a right cylinder, whose axis a is vertical, and the radius of whose base is r; let x denote the depth of flotation. Then we have

---

* Called also depth of *immersion*.

EXAMPLES.

$$v = \pi r^2 a,$$
and
$$v' = \pi r^2 x;$$
$$\therefore \quad x = \frac{v'}{v} a$$
$$= \frac{\rho}{\rho'} a \text{ [from (1)]}.$$

2. Let the body be a right cone, floating with its apex below the surface of the fluid and the axis a vertical. Required the depth of flotation.

Since the volumes of similar cones are proportional to the cubes of their heights, we have, x being the required depth,

$$\frac{v'}{v} = \frac{x^3}{a^3},$$

which in (1) gives,

$$\frac{x^3}{a^3} = \frac{\rho}{\rho'};$$
$$\therefore \quad x = a\sqrt[3]{\frac{\rho}{\rho'}}.$$

3. Let the body be a sphere of radius a, floating in a fluid. Required the depth of flotation.

Here the displaced fluid has the form of a segment of a sphere; hence, calling x the depth, we have, from mensuration,

$$v' = \pi x^2 (a - \tfrac{1}{3}x),$$
and
$$v = \tfrac{4}{3}\pi a^3;$$
$$\therefore \quad \frac{v'}{v} = \frac{3x^2(a - \tfrac{1}{3}x)}{4a^3}$$
$$= \frac{\rho}{\rho'}, \text{ [from (1)]};$$
$$\therefore \quad x^3 - 3ax^2 + 4a^3\frac{\rho}{\rho'} = 0;$$

we have, therefore, to solve a cubic equation in order to find the depth of flotation of the sphere.

4. Let the body be a cylindrical pontoon,* with plane ends, and having its axis horizontal. Required to find the load requisite to sink the pontoon to a given depth.

Let AD be the intersection of the plane of flotation with the end which is a right section. Put A = the area ADK, the plane surface of immersion, and $l =$ AB, the length of the cylinder; and let $W =$ the required load that will sink it to the depth HK. Then, calling ρ' the density of the fluid, we have

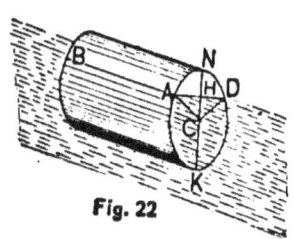

Fig. 22

$$\text{volume of displaced fluid} = Al, \qquad (1)$$

and $\quad$ weight of displaced fluid $= g\rho' Al$;

$$\therefore\ W = g\rho' Al. \qquad (2)$$

A may be found as follows: let $r =$ CK, and $\theta =$ angle ACK; then we have, from mensuration,

$$A = r^2 \left(\pi \cdot \frac{\theta}{180} + \frac{1}{2} \sin 2\theta \right), \qquad (3)$$

which in (2) gives,

$$W = g\rho' r^2 \left(\pi \cdot \frac{\theta}{180} + \frac{1}{2} \sin 2\theta \right) l, \qquad (4)$$

which is the required load.

Cor. 1.—If $\theta = 165°$, we have, from (4),

$$W = \tfrac{1}{4} g\rho' r^2 \left(1\tfrac{1}{3}\pi + 1 \right) l. \qquad (5)$$

---

* Pontoons are portable boats, covered with balks, planks, etc., for forming floating bridges over rivers. They are now usually made of tin, in the shape of a cylinder, with hemispherical ends. (Tate's Mechanical Philosophy.)

Cor. 2.—If the fluid be water, (5) becomes

$$W = \tfrac{1}{4}r^2(\tfrac{1}{3}\pi + 1)\, l\, 62.5 \quad (\text{Art. 10, Cor. 1}). \qquad (6)$$

5. Let the body be a cone floating with its base under the fluid, and the axis a vertical. Find the depth of flotation.

$$Ans.\ \ a - a\sqrt[3]{1 - \frac{\rho}{\rho'}}.$$

6. A man whose weight is 150 lbs. and density 1.1, just floats in water by the help of a quantity of cork. Find the volume of the cork in cubic feet, its density being .24, calling the density of water 1.   *Ans.* $\tfrac{60}{209}$ of a cubic foot.

27. Stability of Equilibrium.—If a floating body is in equilibrium, the centres of gravity and of buoyancy are in the same vertical line (Art. 24, Cor. 1). Imagine the body to be slightly displaced from its position of equilibrium by turning it round through a small angle, so that the axis of flotation shall be inclined to the vertical. If the body on being released return to its original position, its equilibrium is *stable*; if, on the other hand, it fall away from that position, its original position is said to be one of *unstable* equilibrium; when the body neither tends to return to its original position, nor to deviate farther from it, the equilibrium is said to be one of *indifference*.

The investigation of this problem in its utmost extent would lead to very tedious and complex operations, which would clearly be beyond the limits of this treatise; we shall therefore premise the three following hypotheses, in order that we may obtain comparatively simple results:

1. The floating body will be regarded as symmetrical with respect to a vertical plane through its centre of gravity when the whole is at rest, so that we need consider only the problem for the area of a plane section of the body.

2. The displacement will be regarded as very small.

3. The vertical motion of the centre of gravity of the body will be disregarded, as indefinitely small.

Let EDF represent a body which has changed from its upright to its present inclined position, by turning through a small angle; let ABD represent the immersed part of the body before displacement, and HKD that immersed after displacement, and G and O the centres of gravity and of buoyancy before displacement. While the body moves from its upright to its inclined position, its centre of buoyancy moves from O to O', which latter is in the half of the body most immersed, and the wedge-shaped part ACH passes up out of the water, drawing the wedge-shaped part BCK down into it. Let the vertical line through O' meet GO in M.

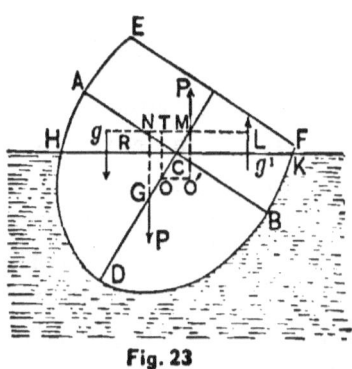

Fig. 23

Now since the buoyant effort is equal to the weight of the whole solid (Art. 24, Sch.), the magnitude of the part immersed will be unaltered; therefore ABD = HKD, and ACH = BCK; also, the buoyant effort P', acting at O' vertically upwards, and the weight P of the solid, acting at G vertically downwards, form a couple which tends to restore the body to its original position when M is *above* G; and, on the contrary, it tends to incline the body farther from its original position when M is *below* G. Hence, the stability of a floating body, a ship, for instance, depends upon the position of the point M, where the vertical line through the centre of buoyancy, in the inclined position of the body, cuts the line connecting the centre of gravity and centre of buoyancy in the upright position of the body.

The position of the point M will in general depend on the extent of displacement. If the displacement be very small,

STABILITY OF EQUILIBRIUM.

i. e., if the angle between GO and the vertical be very small, the point M is called the *metacentre*, and the question of stability is now reduced to the determination of this point. A ship, or any other body, floats with stability when its metacentre lies above its centre of gravity, and without stability when it lies below it; it is in indifferent equilibrium when these two points coincide. Hence the danger of taking the whole cargo out of a ship without putting in ballast at the same time, or of putting the heaviest part of a ship's cargo in the top of the vessel and the lightest in the bottom, or the risk of upsetting when several people stand up at once in a small boat.

One of the most important problems in naval architecture is to secure the ascendancy, under all circumstances, of the metacentre above the centre of gravity. This is done by a proper form of the midship sections, so as to raise the metacentre as much as possible, and by ballasting, so as to lower the centre of gravity.*

The horizontal distance MN, of the metacentre M, from the centre of gravity G of the body, is the arm of the couple whose forces are P and P, the weight of the body and the buoyant effort; and the moment of this couple, which measures the stability of the body, is $P \cdot MN$. Let $GM = c$, and the angle OMO', through which the body rolls, $= \theta$, and denote the measure of the stability by S; then we have

$$S = P \cdot MN = Pc \sin \theta ; \qquad (1)$$

therefore, *the stability of a body, in general, varies as its weight, as the distance of its metacentre from its centre of gravity, and as the angle of inclination; and hence, in the same body, for a given inclination, it depends only upon the distance of its metacentre from its centre of gravity.*

---

* Besant's Hydrostatics, p. 58.

28. The Position of the Metacentre; the Measure of the Stability.

—Since the stability of a body depends principally upon the distance of the metacentre from the centre of gravity of the body, it becomes important to determine the position of the metacentre.

Let A = the cross-section ABD = HKD (Fig. 23) of the immersed part of the body (Art. 27), and A_1 = the cross-section ACH = BCK; let g and g' be the centres of gravity of ACH and BCK; let a = the horizontal distance RL, between these centres of gravity, and s = the horizontal distance between O and O', the centres of buoyancy. Then, taking moments round G, we have,

$$\text{HKD} \times \text{MN} - \text{ACH} \times \text{RN} = \text{ABD} \times \text{NT} + \text{BCK} \times \text{NL};$$

or, $\quad A(\text{MN} - \text{NT}) = A_1(\text{RN} + \text{NL});$

$$\therefore As = A_1 a;$$

or, $\quad s = \dfrac{A_1}{A} a;$

and $\quad \text{OM} = \dfrac{\text{OO}'}{\sin \theta} = \dfrac{A_1 a}{A \sin \theta},$

which is the height of the metacentre above the centre of buoyancy.

Let $\text{GO} = e$; then

$$c = \text{GM} = e + \dfrac{A_1 a}{A \sin \theta}, \qquad (1)$$

which gives the height of the metacentre above the centre of gravity.

Substituting this value of c in (1) of Art. 27, we get

$$S = P\left(\dfrac{A_1 a}{A} + e \sin \theta\right), \qquad (2)$$

which is the measure of the stability.

If the point O were below G, e would be negative and (2) would be

$$S = P\left(\frac{A_1 a}{A} - e \sin \theta\right). \qquad (3)$$

Hence, in general, we have

$$S = P\left(\frac{A_1 a}{A} \pm e \sin \theta\right), \qquad (4)$$

the upper or lower sign being used according as the centre of buoyancy is above or below the centre of gravity.

COR. 1.—If the displacement be small, the cross-sections ACH and BCK can be treated as isosceles triangles, and $\sin \theta = \theta$. Denoting the width AB = HK of the body at the plane of flotation by b, we have

$$A_1 = \tfrac{1}{8} b^2 \theta, \quad \text{and} \quad RL = a = \tfrac{2}{3} b,$$

which in (4) gives

$$S = P\left(\frac{b^3}{12A} \pm e\right)\theta. \qquad (5)$$

COR. 2.—When the centre of buoyancy is above the centre of gravity of the body, the stability is positive, as also in the case when the centre of buoyancy is below the centre of gravity while e is less than $\dfrac{b^3}{12A}$; in this case the equilibrium is that of *stability*.

If e is greater than $\dfrac{b^3}{12A}$, and the centre of buoyancy is below the centre of gravity of the body, the stability is negative, or the equilibrium is that of *instability*.

If e is negative and equal to $\dfrac{b^3}{12A}$, the stability is zero, and the equilibrium is that of *indifference*.

That is, the centre of buoyancy may be *below* the centre of gravity and yet the stability be positive, so long as e does not exceed $\frac{b^3}{12A}$, which term is always the distance between the metacentre and the centre of buoyancy.

If the centre of gravity of the body coincides with the centre of buoyancy, we have $e = 0$, and (5) becomes

$$S = P \frac{b^3}{12A} \theta. \tag{6}$$

Hence, generally, *the stability is positive, negative, or zero, according as the metacentre is above, below, or coincident with the centre of gravity of the floating body.*

A vertical line O'M through the centre of buoyancy is called a *line of support*.

Cor. 3.—From the above results we see that the stability of a body is greater the broader it is and the lower its centre of gravity is. (See Weisbach's Mechs., Vol. I., p. 750; also Bland's Hydrostatics, p. 120.)

EXAMPLES.

1. Determine the stability of a homogeneous rectangular parallelopiped floating in a fluid.

Let HK be the line of flotation of a vertical section passing through the centre of gravity G; let $b =$ the breadth EF of the section of the parallelopiped, $h =$ the height EC, and $y =$ the depth of immersion AC. Then we have

$$A = by, \text{ and } e = -\tfrac{1}{2}(h - y),$$

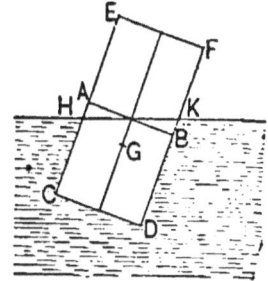

Fig. 24

e being negative since the centre of buoyancy is below the centre of gravity. Substituting in (5), we have

$$S = P\left(\frac{b^3}{12by} - \frac{h}{2} + \frac{y}{2}\right)\theta. \tag{1}$$

Let the density of the material of the parallelopiped be ρ times that of the fluid; then (Art. 25, Cor.),

$$\rho : 1 :: y : h;$$
$$\therefore \quad y = h\rho,$$

which in (1) gives

$$S = \left[\frac{b^2}{12h\rho} - \frac{h}{2}(1 - \rho)\right]P\theta, \tag{2}$$

which is the measure of the stability required.

Cor. 1.—To determine the limits of stability depending upon the dimensions and density of the solid, let $S = 0$, and (2) becomes

$$b^2 - 6h^2\rho(1 - \rho) = 0; \tag{3}$$

or,
$$\frac{b}{h} = \sqrt{6\rho(1 - \rho)}.$$

If $\rho = \frac{1}{2}$, we have

$$\frac{b}{h} = \frac{1}{2}\sqrt{6} = 1.225,$$

and hence in this case the parallelopiped floats in stable, indifferent, or unstable equilibrium, according as the breadth is $>$, $=$, or $<$ 1.225 times the height.

Cor. 2.—Solving (3) for ρ, we get

$$\rho = \frac{1}{2} \pm \frac{1}{2}\sqrt{1 - \frac{2}{3}\frac{b^2}{h^2}},$$

which is real when $\frac{b}{h}$ is $=$ or $< \frac{1}{2}\sqrt{6}$; *i. e.*, when the ratio of the breadth of the solid to the height is equal to, or less than $\frac{1}{2}\sqrt{6}$, two values may be assigned to the density of the solid which will cause it to float in indifferent equilibrium.

If, for instance, $b = h$, we have

$$\rho = \tfrac{1}{2} \pm \tfrac{1}{2}\sqrt{1 - \tfrac{2}{3}} = 0.78868 \text{ or } 0.21132.$$

COR. 3.—When $\dfrac{b}{h} > \frac{1}{2}\sqrt{6}$, the value of ρ is imaginary, *i. e.*, if the ratio of the breadth of the solid to the height is greater than $\frac{1}{2}\sqrt{6}$, no value can be given to the density which will cause the stability to vanish. In this case the solid, placed with EF horizontal, must in all cases continue to float permanently in that position, whatever may be the density, providing it is always less than that of the fluid.

COR. 4.—The term $\dfrac{b^2}{12y}$ in (1), or $\dfrac{b^2}{12h\rho}$ in (2), is the distance between the centre of buoyancy and the metacentre.

2. Determine the angle of inclination θ, in order that the parallelopiped EFDC may be in a position of indifferent equilibrium.

Let $b =$ the breadth EF of the section of the parallelopiped, $y =$ the depth of immersion $AC = BD$, and $\theta =$ angle AOH. Then

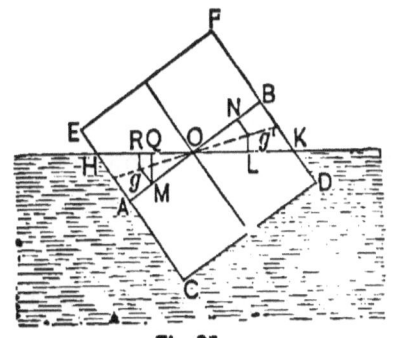

Fig. 25

$A =$ ABDC $=$ HKDC
$= by$, (1)

$A_1 =$ AOH $=$ BOK.

But AO $=$ OB $= \tfrac{1}{2}b$, and AH $=$ BK $= \tfrac{1}{2}b \tan \theta$; therefore

$$A_1 = \tfrac{1}{8}b^2 \tan \theta. \quad (2)$$

EXAMPLES.

Let g and g' be the centres of gravity of the triangles AOH and BOK; draw Mg parallel to AH, and gR and MQ perpendicular to HO. Then

$$Mg = \tfrac{1}{6} b \tan \theta, \quad \text{and} \quad OM = \tfrac{1}{3} b.$$

Therefore, the horizontal distance of the centre of gravity g from the centre O

$$= OR = OM \cos \theta + Mg \sin \theta$$
$$= \tfrac{1}{3} b \cos \theta + \tfrac{1}{6} b \tan \theta \sin \theta;$$

and therefore, for $a = RL = 2OR$, we have

$$a = \tfrac{2}{3} b \cos \theta + \tfrac{1}{3} b \tan \theta \sin \theta. \tag{3}$$

Substituting (1), (2), and (3), in (3) of Art. 28, and putting $S = 0$ for indifferent equilibrium, we get

$$\frac{\tfrac{1}{6} b^2 \tan \theta \, (\tfrac{2}{3} b \cos \theta + \tfrac{1}{3} b \tan \theta \sin \theta)}{by} - e \sin \theta = 0;$$

or, $\qquad [(2 + \tan^2 \theta) \, b^2 - 24 ey] \sin \theta = 0.$

$$\therefore \quad \sin \theta = 0, \tag{4}$$

and $\qquad \tan \theta = \dfrac{1}{b} \sqrt{24ey - 2b^2}. \tag{5}$

The angle $\theta = 0$, in (4), is applicable to the body when in an upright position, and that given in (5) is applicable to the body when floating in an inclined position, and is possible only when b is $=$ or $< 2\sqrt{3ey}$.

COR.—Let $h =$ the height EC, and $\rho =$ the density of the body, the density of the fluid being unity, then we have

$$y = h\rho, \quad \text{and} \quad e = \frac{h}{2}(1 - \rho),$$

which in (5) gives

$$\tan \theta = \frac{1}{b} \sqrt{12h^2 (1 - \rho) \rho - 2b^2}. \tag{6}$$

Hence, when $\dfrac{b}{h} < \sqrt{6\rho(1-\rho)}$, the parallelopiped will float in an inclined position in indifferent equilibrium, the inclination being given by (6).

When $\dfrac{b}{h} > \sqrt{6\rho(1-\rho)}$, the value of $\tan\theta$ is imaginary, *i. e.*, if the ratio of the breadth to the height is greater than $\sqrt{6\rho(1-\rho)}$, no value can be found for the inclination which will cause the stability to vanish. (Compare with last example.)

3. If the breadth of the parallelopiped is equal to its height, and if $\rho = \tfrac{1}{2}$, find the inclination θ, that the parallelopiped may float in indifferent equilibrium.

Ans. $\theta = 45°$

29. Specific Gravity.—*The specific gravity of a body is the ratio of its weight to the weight of an equal volume of some other body taken as the standard of comparison.*

The *density* of a body has been defined (Anal. Mechs., Art. 11), to be the ratio of the mass of the body to the mass of an equal volume of some other body taken as the standard; and since the weights of bodies are proportional to their masses, it follows that the ratio of the weights of two bodies is equal to the ratio of their masses. Hence, the measure of the specific gravity of a body is the same as that of its density, provided that both be referred to the same standard substance.

Thus, let S, W, V, and ρ be the specific gravity, weight, volume, and density, respectively, of one body, and S_1, W_1, V_1, and ρ_1 the same of another body; then we have

$$\frac{W}{W_1} = \frac{g\rho V}{g\rho_1 V_1} = \frac{\rho V}{\rho_1 V_1}, \qquad (1)$$

and making the volumes equal, we have

$$\frac{S}{S_1} = \frac{W}{W_1} = \frac{\rho}{\rho_1};\qquad(2)$$

that is, *the ratio of the specific gravities of two bodies is equal to that of their densities.*

Now suppose the body whose weight is W_1 to be assumed as the standard for specific gravity; then will S_1 be unity, and (2) will become

$$S = \frac{W}{W_1} = \frac{\rho}{\rho_1}.\qquad(3)$$

Also, if the same body be assumed as the standard of density, ρ_1 will be unity, and (3) will become

$$S = \frac{W}{W_1} = \rho.\qquad(4)$$

Hence, *the measure of the specific gravity of a body is the same as that of its density, i. e., the numbers S and ρ are identical, when both specific gravity and density are referred to the same substance as a standard.*

30. The Standard Temperature.—The standard substance to which specific gravity and density are referred is not necessarily the same, and therefore S and ρ will in general be different numbers. In practice, it is usual to adopt water as the standard in determining the specific gravities of solids and incompressible fluids; and for the purpose of rendering the comparison more exact, the water must first be deprived by distillation of any impurities which it may contain.

The dimensions of all bodies being more or less changed by changes of temperature, it becomes necessary to adopt a standard temperature at which experiments for determining specific gravities must be performed. The English * usually take for this purpose the temperature of 60° Fahrenheit, it being easily obtained at all times, and the tables of specific gravities are usually given with reference to distilled water at this temperature as the standard. When the experiment cannot be performed at the standard temperature, the result obtained must be reduced to what it would be at this temperature, *i. e.*, the apparent specific gravity, as obtained by means of water when not at the standard temperature, must be reduced to what it would have been if the water had been at the standard temperature.

Thus, let ρ be the density of any solid, S_1 its apparent specific gravity as obtained by water when not at the standard temperature, and ρ_1 the corresponding density of the water; and let S be the true specific gravity of the body as determined by water at a standard temperature, the corresponding density of the water being ρ_2. Then, from (3) of Art. 29, we have

$$S_1 = \frac{\rho}{\rho_1}, \quad \text{and} \quad S = \frac{\rho}{\rho_2};$$
$$\therefore \frac{S}{S_1} = \frac{\rho_1}{\rho_2}. \tag{1}$$

Calling the density of the standard temperature unity, (1) becomes

$$S = S_1 \rho_1. \tag{2}$$

That is, *the specific gravity of a body as determined at the standard temperature of the water is equal to its specific gravity determined at any other temperature, multiplied by the density of the water at this temperature, the density of the water at the standard temperature being regarded as unity.*

SCH.—In the cases that occur most frequently in practice, such nicety is unnecessary, and the experiment may be

---

* The French usually take the temperature at which water has its maximum of density, which is 39°.4 F.

performed with water at any temperature; but the temperature must be noted and a correction applied for it which depends upon the density of water at the experimental temperature.*

The weight of a cubic foot of distilled water at the standard temperature is 1000 ozs. = 62½ lbs.; hence we find the weight of a cubic foot of any substance in ounces or pounds by multiplying its specific gravity by 1000 or 62½.

It appears, therefore, that by means of the specific gravities of homogeneous bodies, their weights may be determined without actually weighing them, provided their volumes are known; and conversely, however irregular the shape of bodies may be, if their weights and specific gravities are known, their volumes may be determined, viz., by dividing the weight by the specific gravity.

The specific gravities of gases and vapors are usually determined by referring them to atmospheric air at the same temperature and under the same pressure as the gases themselves.

31. Methods of Finding Specific Gravity.

—The law of the buoyant effort, or upward pressure, of water can be made use of to determine the specific gravities of bodies; for, if a body be immersed in a fluid, it loses as much of its weight as is equal to the weight of the fluid it displaces (Art. 24); *i. e.*, if it be wholly immersed, its loss of weight is equal to the weight of its volume of the fluid.

Thus, if a sphere of lead, whose weight is 11 lbs., were found to weigh but 10 lbs. when immersed in water, we should conclude that the weight of an equal volume of water would be one pound, and therefore that the lead weighed 11 times as much as its volume of water, and hence that the specific gravity of lead was 11; and so for any other substance.

---

* Renwick's Mechs., p. 334.

(1) *To find the specific gravity of a solid heavier than water.*

Let $w =$ the weight of the solid in air, $w' =$ its weight in water, and $S =$ the specific gravity of the solid, that of water being 1; then $w - w'$ is the weight lost by the solid, which is also the weight of the water displaced by the solid (Art. 24); therefore w and $w - w'$ are the weights of equal volumes of the solid and water. Hence we have

$$S = \frac{w}{w - w'}. \qquad (1)$$

Hence, to find the specific gravity of a solid heavier than water, we have the following rule: *Divide its weight by its loss of weight in water.*

(2) *To find the specific gravity of a solid lighter than water.*

Since the solid is lighter than water, it will not descend in the water by its own weight; it must therefore be attached to a heavy body of sufficient size and weight to make the two together sink in the water.

Let $w =$ the weight of the solid in air,
$x =$ the weight in air of the heavy body attached to it,
$x' =$ the weight in the water of the heavy body,
$w' =$ the weight in the water of the two together.

Then $w + x - w' =$ the weight of water displaced by the two together.
$x - x' =$ the weight of water displaced by the heavy body.

Hence, $w + x' - w' =$ the weight of water displaced by the solid, and therefore

EXAMPLES.

$$S = \frac{w}{w + x' - w'}. \qquad (2)$$

Hence, *add the difference between the weights of the heavy body and the two together in the water to the weight of the solid in air, and divide the weight of the solid in air by this sum.*

(3) *To find the specific gravity of a liquid.*

Take a solid which is specifically heavier than either the liquid or water, and let it be weighed in both; then the loss of weight in the two cases will be the respective weights of equal volumes of the liquid and of water; therefore, *the loss of weight in the liquid, divided by the loss of weight in the water, will give the specific gravity of the liquid.*

Let w = the weight of the solid in air, w' = its weight in the liquid whose specific gravity is to be determined, and w_1 = its weight in water; then $w - w'$ and $w - w_1$ are the respective weights of equal volumes of the liquid and of water; therefore

$$S = \frac{w - w'}{w - w_1}. \qquad (3)$$

Otherwise thus: Let w = the weight of an empty flask, w' = its weight when filled with the liquid, and w_1 = its weight when filled with water; then $w' - w$ and $w_1 - w$ are the respective weights of equal volumes of the liquid and of water; therefore

$$S = \frac{w' - w}{w_1 - w}. \qquad (4)$$

EXAMPLES.

1. A cubical iceberg is 100 ft. above the level of the sea, its sides being vertical. Given the specific gravity of sea-water = 1.0263, and of ice = 0.9214 at the temperature of 32°, to find its dimensions.

Let $x =$ the length of one side,

$x - 100 =$ the length of the piece under water;

then we have (Art. 25, Cor.),

$$x^3 : x^3 - 100x^2 :: 1.0263 : 0.9214;$$

$$\therefore \ x : 100 :: 1.0263 : 0.1049;$$

$$\therefore \ x = 978.3 \text{ ft.,}$$

and
$$x^3 = 936,302,451.687 \text{ cu. ft.}$$

2. A piece of limestone, whose weight is 256.34 lbs., weighs in water 159.13 lbs. Find its specific gravity.
Ans. 2.637.

3. Find the specific gravity of a piece of cork whose weight is 20 grains. To sink it, we attach a brass weight which, when immersed in the water, weighs 87.22 grains; the weight of the compound body when immersed is 23.89 grains. *Ans.* 0.24.

4. A solid weighing 25 lbs., weighs 16 lbs. in a liquid A, and 18 lbs. in a liquid B. Compare the specific gravities of A and B. *Ans.* 9 : 7.

32. Specific Gravity of a Solid broken into Fragments.—Put the broken pieces into a flask, fill the flask with water, and let its weight be then w''; let w be the weight of the solid in air, and w' the weight of the flask when filled with water. Then

$w'' - w' =$ weight of solid pieces — wt. of water they displace

$= w -$ weight of water displaced;

therefore $w + w' - w'' =$ weight of water displaced;

$$\therefore S = \frac{w}{w + w' - w''}. \tag{1}$$

33. Specific Gravity of Air.

Take a large flask which can be completely closed by a stop-cock, and weigh it when filled with air; withdraw the air by means of an air-pump and weigh the flask again; finally, fill the flask with water and weigh again. This last weight minus the second will give the weight of the water that filled the flask, and the first weight minus the second will give the weight of an equal volume of air; divide the weight of the air by that of the water; the result will be the specific gravity of air as compared with that of water.

Let w = the weight of the exhausted flask; w', w'' its weights when filled with air and water; then

$w' - w$ = weight of the air contained by the flask,

$w'' - w$ = weight of the water contained by the flask;

therefore, $$S = \frac{w' - w}{w'' - w}. \qquad (1)$$

SCH.—In the same manner the specific gravity of any gas can be obtained. The specific gravity of water at $20°.5$ is about 768 times that of air at $0°$ under the pressure of 29.9 inches of mercury at $0°$.

The atmosphere in which these operations must be performed varies at different times, even during the same day, in respect to temperature, the weight of its column which presses upon the earth, and the quantity of moisture it contains. On these accounts, corrections must be made before the specific gravity of air, or that of any gas exposed to its pressure, can be accurately determined. The discussion of the principles according to which these corrections are made, is given in Chap. III.

34. Specific Gravity of a Mixture.

(1) *When the volumes and specific gravities of the components are given.*

Let v, v', v'', etc., be the volumes of the bodies of which the specific gravities are s, s', s'', etc. Then, since the weight of the volume v is $62.5sv$ (Art. 30, Sch.), and so for the others, the weight of the mixture is

$$62.5\,(sv + s'v' + s''v'' + \text{etc.}) = 62.5\,\Sigma\,(sv);$$

and the volume of the mixture is

$$v + v' + v'' + \text{etc.} = \Sigma\,(v);$$

and therefore, if S be the specific gravity of the mixture, we have

$$62.5\,S\,\Sigma\,(v) = 62.5\,\Sigma\,(sv);$$

$$\therefore\; S = \frac{\Sigma\,(sv)}{\Sigma\,(v)}. \tag{1}$$

If by any chemical action the volume becomes V instead of $\Sigma\,(v)$, the specific gravity will be

$$S = \frac{\Sigma\,(sv)}{V}. \tag{2}$$

(2) *When the weights and specific gravities of the components are given.*

Let w, w', w'', etc., be the weights of the bodies, and s, s', s'', etc., their specific gravities. Then, as before, since $w = 62.5sv$, and so for the others, the volumes are respectively $\dfrac{w}{62.5s}$, $\dfrac{w'}{62.5s'}$, etc., and the whole volume is

$$\frac{w}{62.5s} + \frac{w'}{62.5s'} + \text{etc.} = \frac{1}{62.5}\,\Sigma\left(\frac{w}{s}\right);$$

and the whole weight is

$$w + w' + \text{etc.} = \Sigma\,(w);$$

and therefore, if S be the specific gravity of the mixture, we have

$$62.5 S \frac{1}{62.5} \Sigma \left(\frac{w}{s}\right) = \Sigma(w);$$

$$\therefore \ S = \frac{\Sigma(w)}{\Sigma\left(\frac{w}{s}\right)}. \tag{3}$$

Rem.—Instead of taking 1 lb. as the unit of weight, as we have heretofore done, it is sometimes more convenient to take the weight of a unit of volume of the standard substance as the unit of weight; thus, in the present Article, we might have made $62\frac{1}{2}$ lbs. the unit of weight, and found the weights of the substances in terms of that unit.

35. The Weights of the Components of a Mechanical Mixture.

When the specific gravities of the mixture and its components, and also the weight of the mixture are given, to find the weights of the components.

Let w, w', w'' be the weights of the mixture and its components respectively, s, s', s'' their respective specific gravities; and v, v', v'' their volumes. Then we have

$$w = w' + w'', \tag{1}$$

and also

$$v = v' + v'';$$

and, therefore,

$$\frac{w}{s} = \frac{w'}{s'} + \frac{w''}{s''} \ \text{(Art. 34, Rem.).} \tag{2}$$

Combining (1) and (2), we obtain,

$$w' = w\left(\frac{1}{s} - \frac{1}{s''}\right) \div \left(\frac{1}{s'} - \frac{1}{s''}\right)$$

$$= \frac{(s'' - s)\,s'}{(s'' - s')\,s}\, w, \tag{3}$$

$$w'' = w\left(\frac{1}{s} - \frac{1}{s'}\right) \div \left(\frac{1}{s''} - \frac{1}{s'}\right)$$

$$= \frac{(s' - s)\, s''}{(s' - s'')\, s}\, w. \tag{4}$$

EXAMPLES.

1. If with 78 gallons of spirit, specific gravity 0.92, 22 gallons of water be mixed, what is the specific gravity of the mixture? *Ans.* 0.9376.

2. British standard gold contains 11 parts by weight of pure gold, and 1 part of copper. Required its specific gravity. *Ans.* 17.647.

3. An iron vessel completely filled with mercury weighed 500 lbs., and lost, when weighed in water, 40 lbs. If the specific gravity of the cast iron is 7.2 and that of the mercury is 13.6, find (1) the weight of the empty vessel, and (2) that of the mercury contained in it.
Ans. (1) 49.5 lbs.; (2) 450.2 lbs.

36. The Hydrostatic Balance.—In order to determine the specific gravities of bodies practically and with accuracy, it is necessary to employ certain instruments for weighing. These are the *Hydrostatic Balance* and *Hydrometers.**

The hydrostatic balance is an ordinary balance, having one of the scale-pans smaller than the other, and at a less distance from the beam; attached to the under side of the small scale-pan is a hook, from which may be suspended any body by means of a thin platinum wire, horse-hair, or any delicate thread. The body whose specific gravity is to be found is

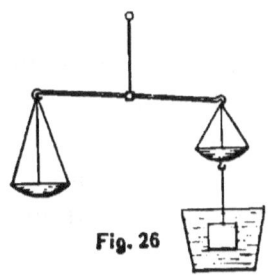

Fig. 26

---

* Sometimes called Areometers.

suspended from the hook, and then its weight is determined. It is then weighed in water, and thus its loss of weight is ascertained, which is the weight of a portion of water equal in volume to the body.

37. The Common Hydrometer.—The name *hydrometer* is given to a class of instruments used for determining the specific gravities of liquids by observing either the depths to which they sink in the liquids or the weights required to make them sink to a given depth. These instruments depend upon the principle that the weight of a floating body is equal to the weight of the fluid which it displaces.

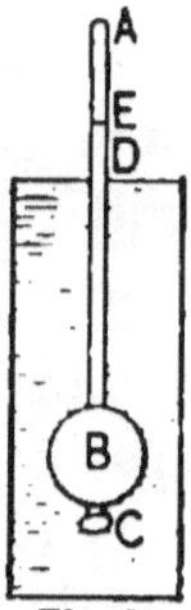

The *common hydrometer* is usually made of glass, and consists of a straight stem ending in two hollow spheres, B and C, the lower one being loaded so as to keep the instrument in a vertical position when floating in the liquid. There are no weights used with the instrument; but the stem is graduated, so as to enable the operator to ascertain the specific gravity of a liquid by the depth to which the instrument sinks in it.

Fig. 27

Let k = the area of a section of the stem, v = the volume, and w = the weight of the hydrometer. When the hydrometer floats in a liquid whose specific gravity is s, let the level D of the stem be in the surface; and when it floats in a liquid whose specific gravity is s', let the level E be in the surface. Then (Art. 34, Rem.) we have for the weights of the liquid displaced in the first and second cases, respectively,

$$w = s(v - k \cdot AD),$$

$$w = s'(v - k \cdot AE);$$

but the weight of the liquid displaced in each case is the same, since each is equal to the weight of the instrument.

$$\therefore \frac{s}{s'} = \frac{v - k \cdot AE}{v - k \cdot AD}, \qquad (1)$$

which gives the ratio of the specific gravities of the two liquids.

Cor.—If the second liquid be the standard, $s' = 1$, and s, the specific gravity of the first liquid, is given in (1).

38. Sikes's Hydrometer.*—This instrument differs from the common hydrometer in the shape of the stem, which is a flat bar and very thin, so that it is exceedingly sensitive. It is generally constructed of brass, and is accompanied by a series of small weights F, which can be slipped over the stem above C, so as to rest on C.

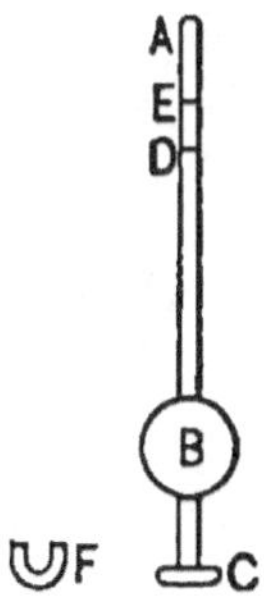

The weights are used to compensate for the great sensitiveness of the instrument, which, without the weights, would render it applicable only to liquids of very nearly the same density.

Let $k =$ the area of a section of the stem, $v =$ the volume, and $w =$ the weight of the hydrometer. When the instrument floats in a liquid whose specific gravity is s, let $w' =$ the weight on C so that the level D of the stem shall be in the surface; and when it floats in a liquid whose specific gravity is s', let $w'' =$ the weight on C so that the level E shall be in the surface; and let v' and v'' be the volumes of w' and w''. Then (Art. 34, Rem.) we have for the weights of the liquid displaced in the first and second cases, respectively,

$$w + w' = s(v + v' - k \cdot AD),$$
$$w + w'' = s'(v + v'' - k \cdot AE);$$
$$\therefore \frac{s}{s'} = \frac{w + w'}{w + w''} \cdot \frac{v + v'' - k \cdot AE}{v + v' - k \cdot AD}. \qquad (1)$$

Fig. 28

---

* Besant's Hydrostatics, p. 127.

Cor.—If the second liquid be the standard, $s' = 1$, and s, the specific gravity of the first liquid, is given in (1).

39. Nicholson's Hydrometer.—The two hydrometers just described are used for obtaining the specific gravities of liquids. *Nicholson's hydrometer* is so contrived as to determine the specific gravity of solids as well as liquids.

It consists of a hollow metallic vessel C, generally of brass, terminated above by a very thin stem, which is often a steel wire, bearing a small dish A, and carrying at its lower end a heavy cup D; on the stem connecting A and C, a well-defined mark B is made.

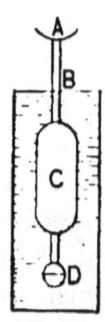

Fig. 29

(1) *To determine the specific gravity of a liquid.*

Let w be the weight of the hydrometer, w' the weight which must be placed in the dish A, in order to sink the stem to the point B in a liquid whose specific gravity is s, and w'' the weight which must be placed in the dish A, to sink the stem to the same point B in a liquid whose specific gravity is s'. Then we have for the weights of the liquid displaced in the first and second cases, respectively,

$$w + w' \quad \text{and} \quad w + w'' :$$

and since the volumes displaced are the same in both cases, the specific gravities are as the weights (Art. 29),

$$\therefore \frac{s}{s'} = \frac{w + w'}{w + w''}. \tag{1}$$

Calling the second liquid the standard, $s' = 1$, and (1) becomes

$$s = \frac{w + w'}{w + w''}, \tag{2}$$

which is the specific gravity required.

(2) *To determine the specific gravity of a solid.*

Let w be the weight which must be placed in the dish A, to sink the stem to the point B in a liquid whose specific gravity is s.

Put the solid in the dish A, and let w' be the weight which must be added to the solid to sink the stem to the point B in the same liquid.

Then put the solid in the lower dish D, and let w'' be the weight required in the upper dish A to sink the stem to the point B in the same liquid.

Hence, the weight of the solid $= w - w'$, and its weight in the liquid $= w - w''$.

Therefore the weight lost, which is the weight of the liquid displaced by the solid $= w'' - w'$. Hence, denoting by S the specific gravity of the solid, we have

$$\frac{S}{s} = \frac{w - w'}{w'' - w'}. \qquad (3)$$

If the liquid is the standard, $s = 1$, and (3) becomes

$$S = \frac{w - w'}{w'' - w'}, \qquad (4)$$

which is the specific gravity required.

EXAMPLES.

1. If an iceberg whose density is 0.918 float in a liquid whose density is 1.028, what is the ratio of the part submerged to that which is above water? *Ans.* 8.3 : 1.

2. How much of its weight will 112 lbs. of iron lose, if immersed in water, the density of iron being 7.25 times that of water? *Ans.* 15.448 lbs.

3. If 20 lbs. of cork be immersed in water, with what force will it rise towards the surface, its density being 0.24 times that of water? *Ans.* $63\frac{1}{3}$ lbs.

EXAMPLES. 81

4. If a piece of wood whose vertical height is 2 ft. be placed in the Dead Sea, how many inches will it become submerged, the densities of the wood and Dead Sea water being .53 and 1.24 respectively? *Ans.* 10.26 ins.

5. Find the depth to which a rectangular block will sink in water, the depth of the block being a feet, and the weight of each cubic foot of it being w lbs. *Ans.* $\dfrac{aw}{62.5}$.

6. A barge of a rectangular shape is l ft. long, b ft. broad, and a ft. deep, outside measure. The thickness of the planking is e ft., and the weight of a cubic foot of the timber is w lbs. To what depth will the barge sink when loaded with W lbs.?

Ans. $\dfrac{w[abl - (a-e)(b-2e)(l-2e)] + W}{62.5bl}$.

7. A cylindrical piece of wood, weight W, floats in water with its axis vertical and immersed to a depth h. Find how much it will be depressed by placing a weight w on the top of it. *Ans.* $\dfrac{w}{W}h$.

8. An isosceles triangle floats in water with its base horizontal. Find the position of equilibrium when the base is above the surface, its height being h and its density being $\frac{2}{3}$ that of water. *Ans.* $\dfrac{h}{3}\sqrt{6}$.

9. A rectangular barge, l ft. long, b ft. broad, and a ft. deep, outside measure, sinks to $\frac{1}{5}$ its whole depth when unloaded. Required its weight in lbs. *Ans.* $12.5abl$.

10. If a rectangular barge sinks to $\frac{1}{4}$ of its whole depth when unloaded, and to $\frac{3}{4}$ of its whole depth when loaded, find the load, the weight of the barge being w. *Ans.* $2w$.

11. The diameter of the base of a right cone is $2r$, its altitude is h, and its density is $\frac{2}{3}$ that of water. To what

depth will the cone sink when it floats with its vertex downwards? $\qquad$ Ans. $\dfrac{h}{3}\sqrt[3]{18}$.

12. A hemispherical vessel, whose weight is w, floats upon a fluid with $\frac{1}{3}$ of its radius below the surface. What weight must be put into the vessel so that it may float with $\frac{2}{3}$ of its radius below the surface? Ans. $\frac{8}{5}w$.

13. Let the pontoon in Ex. 4, Art. 26, be a cylinder, length l, with hemispherical ends, radius r; to find the load requisite to sink the pontoon to a given depth a.

Ans. $[Al + \pi a^2(r - \tfrac{1}{3}a)]\,62.5$,
where $A =$ the area ADK (Fig. 22).

14. Required the thickness of a hollow globe of copper whose density is 9 times that of water, so that it may just float when wholly immersed in water, r being the exterior radius. Ans. $r\left(1 - \tfrac{2}{3}\sqrt[3]{3}\right)$.

15. A cubical box, the volume of which is one cubic foot, is three-fourths filled with water, and a leaden ball, the volume of which is 72 cubic inches, is lowered into the water by a string. It is required to find the increase of pressure (1) on the base and (2) on a side of the box.

Ans. (1) $41\frac{2}{3}$ oz.; (2) $32+$ oz.

16. If the height of the parallelopiped in Ex. 2, Art. 28, is 0.9 of the breadth, and if $\rho = \tfrac{1}{2}$, find the inclination θ that the parallelopiped may float in indifferent equilibrium.

Ans. $\theta = 33°\ 15'$.

17. What is the weight of a cube of gold whose side is 3 ins., its specific gravity being 19.35? Ans. 18.896 lbs.

18. What is the volume of a piece of platinum whose weight is 10 lbs., its specific gravity being 22.06?

Ans. 12.533 cu. ins.

19. A piece of lead, whose weight is 511.65 grs., weighs in water 466.57 grs. Required its specific gravity.

Ans. 11.35.

20. A sovereign, whose weight is 123.02 grs., weighs in water 116.02 grs. Required its specific gravity.
Ans. 17.574.

21. Find the specific gravity of a piece of wood whose weight is 50 grs. To sink it we attach a brass weight which, when immersed in the water, weighs 87.22 grs.; the weight of the compound body when immersed is 42.88.
Ans. 0.53.

22. A piece of wood weighs 4 lbs. in air and a piece of lead weighs 4 lbs. in water; the lead and wood together weigh 3 lbs. in water. Find the specific gravity of the wood. *Ans.* 0.8.

23. A body immersed in water is balanced by a weight P, to which it is attached by a string passing over a fixed pulley; when half immersed, it is balanced in the same way by a weight $2P$. Find the specific gravity of the body.
Ans. $\frac{3}{2}$.

24. Find the weight of a cubical block of stone whose side is 4 ft., and specific gravity $1\frac{1}{4}$. *Ans.* 80000 oz.

25. A body weighing 20 grs. has a specific gravity of $2\frac{1}{2}$. Required its weight in water. *Ans.* 12 grs.

26. An island of ice rises 30 ft. out of the water, and its upper surface contains $\frac{3}{4}$ of an acre. Supposing the mass to be cylindrical, required (1) its weight, and (2) depth below the water, the specific gravity of sea-water being 1.0263, and that of ice .92. *Ans.* (1) 242900 tons; (2) 259.64 ft.

27. A piece of wood weighs 12 lbs., and when attached to 22 lbs. of lead and immersed in water, the whole weighs 8 lbs. The specific gravity of lead being 11, required that of the wood. *Ans.* $\frac{1}{2}$.

28. A solid which is lighter than water weighs 5 lbs., and when it is attached to a piece of metal, the whole weighs 7 lbs. in water. The weight of the metal in water

being 9 lbs., compare the specific gravities of the solid and of water. *Ans.* 5 : 7.

29. A piece of wood which weighs 57 lbs. in vacuo, is attached to a bar of silver weighing 42 lbs., and the two together weigh 38 lbs. in water. Find the specific gravity of the wood, that of water being 1, and that of silver 10.5.
Ans. 1.

30. Equal weights of two fluids, whose specific gravities are s and $2s$, are mixed together, and one-third of the whole volume is lost. Find the specific gravity of the resulting fluid. *Ans.* $2s$.

31. Two fluids of equal volume, and of specific gravities s and $2s$, lose $\frac{1}{4}$ of their whole volume when mixed together. Find the specific gravity of the mixture. *Ans.* $2s$.

32. A cylinder floats vertically in a fluid with 8 ft. of its length above the fluid; find the whole length of the cylinder, the specific gravity of the fluid being three times that of the cylinder. *Ans.* 12 ft.

33. A body floats in one fluid with $\frac{3}{4}$ of its volume immersed, and in another with $\frac{4}{5}$ immersed. Compare the specific gravities of the two fluids. *Ans.* 15 : 16.

34. A block of wood, the volume of which is 4 cubic feet, floats half immersed in water. Find the volume of a piece of metal, the specific gravity of which is 7 times that of the wood, which, when attached to the lower portion of the wood, will just cause it to sink. *Ans.* $\frac{2}{3}$ of a cubic foot.

35. A cone, whose specific gravity is $\frac{1}{2}$, floats on the water with its axis vertical, (1) with its vertex downwards and (2) with its vertex upwards. What part of the axis is immersed in each case? *Ans.* (1) $\frac{1}{2}$; (2) 0.0436.

36. A cone, whose specific gravity is $\frac{1}{2}$, floats with its axis vertical. Compare the portions of the axis immersed, (1) when the vertex is upwards, (2) when it is downwards.
Ans. $\sqrt[3]{2} - 1 : 1$.

37. A block of ice, the volume of which is a cubic yard, is observed to float with $\frac{2}{5}$ of its volume above the surface, and a small piece of granite is seen embedded in the ice. Find the size of the stone, the specific gravities of ice and granite being respectively .918 and 2.65.

Ans. $\frac{41}{860}$ of a yard.

38. A cylindrical glass cup weighs 8 ozs., its external radius is $1\frac{1}{4}$ ins., and its height $4\frac{1}{2}$ ins. If it be allowed to float in water with its axis vertical, find what additional weight must be placed in it, in order that it may sink.

Ans. $\left(\frac{375\pi}{64} - 8\right)$ oz.

39. Find the position of equilibrium of a cone, floating with its axis vertical and vertex upwards, in a fluid of which the density bears to the density of the cone the ratio 27 : 19.

Ans. $\frac{1}{3}$ of the axis is immersed.

40. The whole volume of a hydrometer is 5 cu. ins., and its stem is $\frac{1}{4}$ of an inch in diameter; the hydrometer floats in a liquid A, with one inch of the stem above the surface, and in a liquid B with two ins. above the surface. Compare the specific gravities of A and B.

Ans. $1280 - \pi : 1280 - 2\pi$.

41. What volume of cork, specific gravity .24, must be attached to 6 lbs. of iron, specific gravity 7.6, in order that the whole may just float in water?

Ans. $\frac{198}{1805}$ of a cubic foot.

42. If a piece of metal weigh in vacuum 200 grs. more than in water, and 160 grs. more than in spirit, what is the specific gravity of spirit? *Ans.* $\frac{4}{5}$.

43. A piece of metal whose weight in water is 15 ozs., is attached to a piece of wood, which weighs 20 ozs. in vacuum, and the weight of the two in water is 10 ozs. Find the specific gravity of the wood. *Ans.* $\frac{4}{5}$.

44. A crystal of salt weighs 6.3 grs. in air; when covered with wax, the specific gravity of which is .96, the whole weighs 8.22 grs. in air and 3.02 in water. Find the specific gravity of salt. *Ans.* 1.9 nearly.

45. A Nicholson's Hydrometer weighs 6 ozs., and it is requisite to place weights of 1 oz. and $1\frac{1}{2}$ ozs. in the upper cup to sink the instrument to the same point in two different liquids. Compare the specific gravities of the liquids.
Ans. 4 : 5.

46. A diamond ring weighs $69\frac{1}{2}$ grs., and $64\frac{1}{4}$ grs. in water. The specific gravity of gold being $16\frac{1}{2}$, and that of diamond $3\frac{1}{2}$, what is the weight of the diamond?
Ans. $3\frac{1}{2}$ grs.

47. A body A weighs 10 grs. in water, and a body B weighs 14 grs. in air, and A and B together weigh 7 grs. in water. The specific gravity of air being .0013, required (1) the specific gravity of B, and (2) the number of grs. of water equal to it in volume.
Ans. (1) .8237; (2) 17.023 grs.

48. A compound of gold and silver, weighing 10 lbs., has a specific gravity of 14, that of gold being 19.3, and that of silver being 10.5. Required the weights of the gold and the silver in the compound.
Ans. Gold = 5.483 lbs.; silver = 4.517 lbs.

49. A diamond ring weighs 65 grs. in air and 60 in water. Find the weight of the diamond, if the specific gravity of gold is 17.5, and that of the diamond $3\frac{1}{2}$. *Ans.* 6.875 grs.

50. The crown made for Hiero, King of Syracuse (Art. 24, note), with equal weights of gold and silver, were all weighed in water; the crown lost $\frac{1}{14}$ of its weight, the gold lost $\frac{1}{17}$ of its weight, and the silver lost $\frac{2}{21}$ of its weight. Prove that the gold and silver were mixed in the proportion of 11 : 9.

51. A ring consists of gold, a diamond, and two equal rubies; it weighs 44¼ grs., and in water 38¾ grs.; when one ruby is taken out, it weighs 2 grs. less in water. Find the weight of the diamond, the specific gravity of gold being 16½, of diamond 3½, of ruby 3. *Ans.* 5¼ grs.

52. If the price of pure whiskey be $4 per gallon, and its specific gravity be .75, what should be the price of a mixture of whiskey and water which on gauging is found to be of specific gravity .8? *Ans.* $3.20.

53. How deep will a paraboloid sink in a fluid whose specific gravity is n times that of the solid, the axis being vertical and equal to a, and the vertex upwards?

$$\text{Ans. } a \cdot \frac{\sqrt{n} - \sqrt{n-1}}{\sqrt{n}}.$$

54. A cubic inch of metal, whose specific gravity is m, is formed into a hollow cone, and immersed with its vertex downwards. Determine the ratio of the altitude to the exterior radius of its base, when the surface immersed is a minimum. *Ans.* $\sqrt{2}$.

CHAPTER III.

EQUILIBRIUM AND PRESSURE OF GASES. — ELASTIC FLUIDS.

40. Elasticity of Gases.—The pressure of an elastic fluid is measured exactly in the same way as the pressure of a liquid (Art. 6), and the equality of pressure in every direction, and of transmission of pressure, are equally true of liquids and gases (Arts. 7 and 8). There is, however, this difference between a liquid and a gas: when a liquid is confined in a vessel, no pressure is exerted against the sides except that which is due to the weight of the liquid itself, or that which is transmitted by the liquid from some point on the surface at which an external force is applied; whereas, if a gas be contained in a closed vessel, there is, although modified by the action of gravity, an outward pressure exerted against the sides, which is due to the elasticity of the gas, and which depends upon its volume and temperature.* It is therefore evident that generally a gas cannot have a free surface like a liquid (Art. 11), for such a surface implies that at each point the pressure is nothing, *i. e.*, if it be covered by an envelope everywhere in close contact with it, no pressure is exerted against the envelope. It is also evident that, if a portion of the gas be withdrawn from the vessel, that which remains will not fill the same part of the vessel that it occupied before, as in the case of a liquid, but will expand so as to fill the whole vessel, pressing, but with diminished force, against its sides at every point (Art. 2). From this property of gases, they are called *elastic fluids;* the outward pressure which a gas exerts

---

* If the gas is not confined within a limited space, the effect of its elasticity might be the unlimited expansion and ultimate dispersion of the gas.

against the walls of the vessel enclosing it is called its *elastic force.*

The action of a common syringe will serve to illustrate the elasticity of atmospheric air. If the piston be drawn out, and the open end of the syringe then closed, a considerable effort will be required to force in the piston to more than a small part of the length of its range, and if the syringe be air-tight and strong enough, it will require the application of great power to force the piston down through nearly the whole of its range. This experiment also shows that the pressure increases with the compression, the air within the syringe acting as an elastic cushion. If the piston be let go, after being forced in, it will be driven back, the air within expanding to its original volume.

An inverted glass cylinder, carefully immersed in water, furnishes another simple illustration of the elasticity of air. Holding the cylinder vertical, it may be pressed down in the water without much loss of air, and it will be seen that the surface of the water within the vessel CD is below the surface of the water outside AB. It is evident that the downward pressure of the air within at CD is equal to the upward pressure of the water at the same place, which (Art. 11, Cor. 2) is equal to the pressure on the upper surface AB, increased by the pressure due to the depth of the surface CD below the upper surface; hence the air within, which has a diminished volume, has an increased pressure.

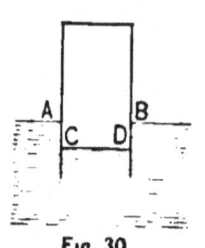

Fig. 30

41. Pressure of the Atmosphere.—If a glass tube* about three feet in length, closed at one end, be filled with mercury, and then, with the finger pressed to the open end

---

* This experiment was first made by Torricelli, and hence is called *Torricelli's Experiment*, and the vacant space above the mercury in the tube is called the *Torricellian Vacuum.*

so as to close it, inverted in a vessel of mercury so as to immerse its open end, it will be found on removing the finger that the mercury in the tube will descend through a certain space, leaving a vacuum at the top of the tube, but resting with its upper surface at a height of about 29 or 30 inches above the surface of the mercury in the vessel. It thus appears that the atmospheric pressure, acting on the surface of the mercury in the vessel, and transmitted (Art. 8), supports the column of mercury in the tube, and hence that the weight of the mercurial column is exactly equal to the weight of the atmospheric column standing on an area equal to that of the internal section of the tube. The weight of this column of mercury then is an exact measure of the atmospheric pressure, or of the elastic force of the atmosphere at any instant.

42. Weight of the Air.—This may be directly proved by weighing a flask filled with air, and afterwards weighing it when the air has been withdrawn by means of an air-pump; the difference of the weights is the weight of the air contained by the flask.

The opinion was long held that air was without weight, or rather, it never occurred to any of the philosophers who preceded Galileo to attribute any influence in natural phenomena to the weight of the air. The fact that air has weight escapes common observation in consequence of its extreme levity compared with solids and liquids, and especially in consequence of its being the medium by which we are continually surrounded. The experiment of weighing air was performed successfully for the first time in 1650, by Otto Guericke, the inventor of the air-pump.*

By means of the weight of air we may account for the fact of atmospheric pressure. The earth is surrounded by a quantity of air, the height of which is limited (see Art.

---

* Deschanel's Natural Philosophy, Part I., p. 141.

72); and if we suppose a cylindrical column extending above any horizontal area to the surface of the atmosphere, the weight of the column of air must be entirely supported by the horizontal area upon which it rests, and the pressure upon the area is therefore equal to the weight of the column of air. The pressure of the air must then diminish as the height above the earth's surface increases; and from experiments in balloons and in mountain ascents, this is found to be the case.

The action of gravity is equivalent to the effect of a compression of the gas, and it is thus seen that the pressure of a gas is in fact caused by its weight, as in the case of a liquid.

Taking π for the pressure of the air at any given place (Art. 11, Cor. 2), and assuming that the density of the air throughout the height z is constant and equal to ρ, the pressure at the height z will be

$$\pi - g\rho z. \qquad (1)$$

Cor.—It may be shown, in the same manner as for air, that any other gas has weight, and that the intrinsic weight is in general different for different gases. Carbonic acid gas, for instance, is heavier than air, and this is illustrated by the fact that it can be poured, as if it were liquid, from one jar to another.

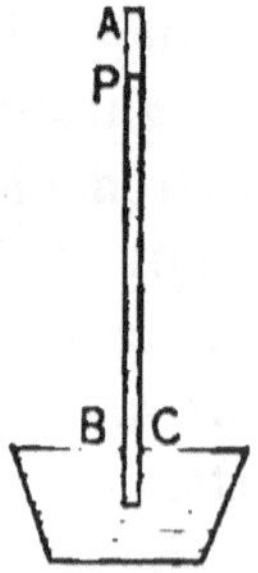

43. The Barometer. — This instrument, which is employed for measuring the pressure of the atmosphere, is, in its simplest form, a straight glass tube AB, about 32 or 33 inches long, containing mercury, and having its lower end immersed in a small cistern of mercury; the end A is hermetically sealed, and there is no air in the branch AB. Since the pressure of a fluid at rest is the same at all points of the same horizontal plane (Art. 10), the pressure at B, in the interior of the tube, is equal to

Fig. 31

the atmospheric pressure on the mercury at C, which is transmitted from the surface of the mercury in the cistern to the interior of the tube; and as there is no pressure on the surface at P, it is clear that the pressure of the air on C is the force which sustains the column of mercury PB.

Let σ be the density of mercury, and π the atmospheric pressure at C; then we have

$$\pi = g\sigma\text{PB}, \qquad (1)$$

and, since g and σ are constant, the height PB may be used as a measure of the atmospheric pressure.

44. The Mean Barometric Height.—The mean height of the barometric column at the level of the sea is found to vary with the latitude, but it is generally between 29½ and 30 inches. The atmosphere is subject to continual changes, some irregular, others periodical. If the density and consequent elastic force of the air be increased, the column of mercury will rise till it reaches a corresponding increase of weight; if, on the contrary, the density of the air diminish, the column will fall till its diminished weight is sufficient to restore the equilibrium. The barometric height is therefore subject to continuous variations; during any one day there is an oscillation in the column, and the mean height for one day is itself subject to an annual oscillation, independently of irregular and rapid oscillations due to high winds and stormy weather. Usually the height of the column is a maximum about 9 A. M.; it then descends until 3 P. M., and again attains a maximum at 9 P. M.*

45. The Water-Barometer.—Mercury possesses two great advantages over other liquids, which has led to its being selected above all others for use in barometric instruments. The first advantage of mercury is that it does not give off vapor at ordinary temperatures. If it did, the space

---

* Besant's Hydrostatics, p. 75.

AP above the mercury would be filled with an elastic vapor, which would press down upon the column, so that its weight would no longer be a measure of the atmospheric pressure, but of the difference between this pressure and the elastic force of the vapor given off. The second advantage is that, on account of the great density of mercury, the height of the column which measures the atmospheric pressure is so small that barometers constructed with it are of a very convenient size. The pressure of the air may be measured by using any kind of liquid. The density of mercury is about 13.595 times that of water,* and therefore, if water were used, it would be necessary to have a tube of great length, since the column of water in the water-barometer would be about $33\frac{1}{4}$ feet.

In order to measure easily and correctly the barometric height, an accurately graduated scale is added, which can be moved along the tube.

Rem.—The instrument above described involves the essential parts of a barometer; it is the province of Physics to give a full description of different kinds of barometers, to explain their use, etc.

46. Manometers.—Barometers are used not only to measure the pressure of the external air, but also to determine the elastic force of gases or vapors which are enclosed in vessels. When thus used, they are called *manometers*. These instruments are filled with mercury, and are either open or closed; in the latter case, there may be air above the column of mercury or there may be a vacuum. The manometer with a vacuum above the column of mercury is like the common barometer. In order to measure with it the elastic force of the gas or vapor, it will be necessary to establish a free connection between the cistern of the barometer and the vessel containing the fluid. This is done by means of a

---

* Enc. Brit., Vol. XVI., p. 33.

tube DE, one end of which E opens into the vessel containing the fluid, and the other end D enters above the level of the mercury B in the cistern. By this means the gas from the vessel flows through the tube ED into the cistern, and presses a column of mercury into the tube AB, the height of which measures the elastic force of the gas or vapor in the vessel.

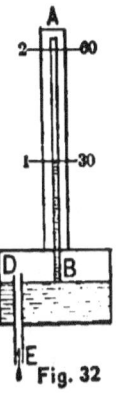

Fig. 32

When the elastic force of the fluid is considerable, it is usual to estimate it as so many atmospheres: for instance, steam, in the boiler of an engine, having a pressure of two atmospheres, signifies that its elastic force would sustain a column of about 60 inches of mercury. If it is said to have a pressure of 6 atmospheres, it means that its elastic force would sustain a column of about 180 inches of mercury; and so on.

47. The Atmospheric Pressure on a Square Inch.
—This may be found at once by observing that it is the weight of a cylindrical column of mercury whose base is a square inch, and whose height is equal to that of the barometric column.

Since the specific gravity of mercury is 13.595, that of water being 1, it follows that the pressure of the air on a square inch, taking 30 inches as the height of the barometer at the sea level,

$$= (30 \times 13.595 \times 62.5 \div 1728) \text{ lbs.}$$
$$= 14.7 \text{ lbs.},$$

and this is called *the pressure of one atmosphere.*

Sch.—This pressure varies from time to time, but is generally between 14½ and 15 lbs. The standard usually adopted where the English system of measure is used is 14.7 lbs. upon the square inch, which corresponds to a column of mercury about 30 (exactly 29.922) inches, and to a column of water about 34 (exactly 33.9) feet high. A pressure of two atmospheres, therefore, would mean a pressure of 29.4 lbs. on each square inch, and a pressure of six atmospheres would mean a pressure of 88.2 lbs. on each square inch. (See Weisbach's Mechs., p. 777.)

EXAMPLES.

1. If the elastic force of a gas is $2\frac{1}{2}$ atmospheres, find its pressure in lbs. on each square inch. *Ans.* 36.75 lbs.

2. If the elastic force of steam in a boiler be $5\frac{1}{2}$ atmospheres, find the pressure on a safety-valve whose area is 5.4 sq. ins. *Ans.* 436.59 lbs.

48. Boyle and Mariotte's Law.*—Gases readily contract into smaller volumes when compressed. When a gas is compressed, its elastic force is increased; and when it is allowed to expand, its elastic force is diminished. The statement of the law which expresses the relation between the pressure and the volume, or the pressure and the density, of gases is the following:

The pressure of a given quantity of air, at a given temperature, varies inversely as its volume, and directly as its density.

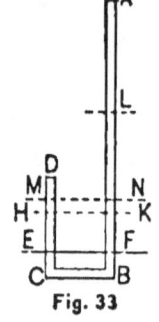

Fig. 33

Let ABCD be a bent glass tube, the shorter branch of which can have its end D closed, and both branches being vertical. Let a little mercury be poured in at A, and let it stand at the same level EF in both branches. Now close the end D; a definite volume of air is thus enclosed in DE under a pressure equal to that of the external air, *i. e.*, the elastic force of the enclosed air DE is equal to the atmospheric pressure exerted on F in

---

* The experimental proof of this law was discovered about the same time in England by the Hon. Robert Boyle, and in France by Mariotte.

the open branch, and is therefore equal to one atmosphere (Art. 47).

Take $DH = \frac{1}{2}DE$, and pour mercury slowly into the tube AB till it stands at H in the shorter branch; in the longer branch it will be found to stand at the height $LK = 30$ inches above HK, *i.e.*, the mercury, rising in the shorter branch, compresses the air which it drives before it, and when the air in the shorter branch is reduced to half its volume, its elastic force or pressure is two atmospheres, since it now sustains not only the atmospheric pressure which is exerted on the surface of the mercury in the open branch, but also the weight of a column of mercury 30 inches high. When mercury is poured into the tube till it rises in the shorter branch to M, where $DM = \frac{1}{3}DE$, it will be found to stand in the longer branch at the height $AN = 60$ inches above MN, *i. e.*, when the air in the shorter branch is reduced to one-third of its volume, its elastic force or pressure is three atmospheres, since it now sustains the atmospheric pressure and the weight of a column of mercury 60 inches in height. In the same way, it may be shown that if the air occupy one-fourth of its original volume DE, it will sustain a pressure of four atmospheres, and so on for any number. Hence, generally, the *pressure* of a quantity of air varies *inversely as its volume*.

When the volume is reduced to one-half, the density is doubled; when reduced to one-third, the density is trebled, and so on; that is, the *volume* varies *inversely as the density*. Hence, the *pressure* varies *directly as the density*.

Let v and v' be the volumes of a given mass of air, p and p' the corresponding pressures, and ρ and ρ' the corresponding densities. Then we have

$$\frac{p}{p'} = \frac{v'}{v} = \frac{\rho}{\rho'}, \qquad (1)$$

$$p = k\rho, \qquad (2)$$

where k is a constant to be determined by experiment.

Rem. 1.—It has been shown by a series of experiments that this law connecting the elastic force and volume of a gas under a constant temperature is sensibly true for air and most gases as far as a pressure of 100 atmospheres.* It is only when the pressures are very great that variations from the law are observed, and even then the departure from the law is but small, especially with those gases which we are not able to condense into liquids. With gases which undergo liquefaction at moderate pressures, the departure from the law is greater, and increases as the state of liquefaction is approached.†

Rem. 2.—In conducting this experiment, care must be taken to have the temperatures the same at the beginning and at the conclusion, as the elastic force of a gas under a given volume is influenced by changes of temperature. For this reason, it is necessary to pour in the mercury gradually, and to allow some time to elapse before the difference of levels is observed, since, whenever a gas is compressed, an elevation of temperature is produced. Therefore, whatever heat is developed by increase of pressure must be allowed to pass off before the volume of gas is observed.

EXAMPLES.

1. Let DE (Fig. 33), be 10 inches; if mercury be poured in until the level in the closed branch stands 3 inches above EF, and in the open branch 15.64 inches, find the elastic force of the air in the closed branch, the barometer standing at 29.5 inches.

Since the levels of the mercury in the two branches stand at 15.64 and 3 inches, the level in the longer branch is 12.64 inches above that in the closed branch; the elastic force of the compressed air, therefore, sustains a column of

---

* Galbraith's Hydrostatics, p. 35.
† Weisbach's Mechs., p. 782; also Twisden's Mechs., p. 229.

mercury 12.64 inches high, together with the atmospheric pressure, which by the barometer is shown to be equal to a column of 29.5 inches; hence the elastic force

= (12.64 + 29.5) inches = 42.14 inches.

2. If the level in the closed branch rise 6.4 inches, find the height to which the level in the open branch should rise, the barometer standing at 30.42 inches, and DE being 10 inches. *Ans.* 60.48 inches.

49. Effect of Heat on Gases.—When a given quantity of air or gas is increased in temperature, it is found that, if the air or gas cannot change its volume, its elastic force is increased; but if the air can expand freely, while its elastic force remains the same, its volume will be increased.

To illustrate this, take an air-tight piston in a vertical cylinder containing air, and let it be in equilibrium, the weight of the piston being supported by the cushion of air beneath it. Raise the temperature of the air in the cylinder by immersing it in hot water; (1) the piston will rise in the cylinder as the volume of the heated air expands; and when the air has reached the temperature of the surrounding water, the piston will cease to ascend, and will remain stationary. But (2) if we suppose that when the heat is applied, the piston is held down so as to keep the air under a constant volume, an effort will be required to prevent the piston from ascending in the tube, which becomes greater in proportion as the air is heated. Hence

(1) *The effect of heat on a given quantity of air, the elastic force remaining constant, is to expand its volume.*

(2) *The effect of heat on a given quantity of air, the volume remaining constant, is to increase its elastic force.*

50. Thermometers.—As a general rule, bodies expand under the action of heat, and contract under the action of cold, and the only method of measuring temperatures is by observing the extent of the expansion or contraction of some known substance. Any body which indicates changes of temperature may be called a *thermometer*.

As the expansions of different substances are not exactly proportional to one another, it is necessary to select some one substance or combination of substances to furnish a standard, and the standard usually adopted for all ordinary temperatures is the apparent expansion of mercury in a graduated glass vessel; for very high temperatures, a *metal* of some kind is the more useful, and for very low temperatures, at which mercury freezes, *alcohol* must be employed.

The mercurial thermometer is formed of a thin glass tube of uniform bore, terminating in a bulb, and having its upper end hermetically sealed. The bulb contains mercury, which also extends partly up the tube, and the space between the mercury and the top of the tube is a vacuum. Since the glass, as well as the mercury, expands with an increase of temperature, the apparent expansion is the difference between the actual expansion and the expansion of the glass. The construction of an accurate mercurial thermometer is an operation of great delicacy.

In *Fahrenheit's Thermometer*, which is chiefly used in England and in this country, the freezing point is marked 32°, and the boiling point 212°. The space, therefore, between these two points is 180°.

In the *Centigrade Thermometer* the freezing point is marked 0°, and the boiling point 100°, the space between being divided into 100°.

In *Reaumur's Thermometer* the freezing point is also marked 0°, but the boiling point is marked 80°.*

---

* The temperature indicated by the boiling point is the same in all.

REM.—Mercury freezes at a temperature of $-40°$ C. or F., and boils at a temperature of about 350° C. or 662° F.; it is therefore necessary, for very high or very low temperatures, to employ other substances.

For very low temperatures, spirit of wine is used; this liquid has never congealed, although a temperature of $-140°$ C. has been observed, which is the lowest temperature yet attained.*

High temperatures are compared by observing the expansion of bars of metal or other solid substances, and instruments called *pyrometers* have been constructed for this purpose.

51. Comparison of the Scales of these Thermometers.—Any degrees of temperature by either thermometer may be converted into the corresponding degrees of the other thermometers; for the space between the fixed points in Fahrenheit's being 180°, in the Centigrade 100°, and in Reaumur's 80°, we have 180° Fahrenheit = 100° Centigrade = 80° Reaumur; and therefore each of Fahrenheit's degrees = $\frac{5}{9}$ of one of Centigrade = $\frac{4}{9}$ of one of Reaumur.

Let F, C, and R be the numbers of degrees marking the same temperature on the respective thermometers; then since the space between the boiling and freezing points must in each case be divided in the same proportion by the mark of any given temperature, we must have

$$\frac{F-32}{180} = \frac{C}{100} = \frac{R}{80};$$

or,
$$\frac{F-32}{9} = \frac{C}{5} = \frac{R}{4}. \qquad (1)$$

REM.—The various scales were formed in the early part of the 18th century—Fahrenheit's in 1714, at Dantzic; Reaumur's in 1731; and the Centigrade somewhat later.†

EXAMPLES.

1. What temperatures on the other two scales are equivalent to the temperature 50° F.?‡ *Ans.* 10° C., or 8° R.

---

* Maxwell on Heat.   † Besant's Hydrostatics, p. 88.

‡ It is usual, in stating temperatures, to indicate the scale referred to by the initials F., C., R.

2. Find (1) what temperature C. is the same as 60° R., and (2) what temperature R. is the same as 45° C.

Ans. (1) 75° C.; (2) 36° R.

52. Expansion of Mercury.—The expansion of mercury is very nearly uniform between 0° and 300°. Experiments show that, for an increase of 1° Centigrade, the expansion of mercury is $\frac{1}{5510}$, or .0001815 of its volume;* hence, if σ_t be the density at a temperature t, and σ_0 the density at a temperature 0°, we have

$$\sigma_0 = \sigma_t (1 + .00018018 t);$$

or, if we put $.00018018 = \theta$, we have

$$\sigma_0 = \sigma_t (1 + \theta t), \qquad (1)$$

which, in (1) of Art. 43, gives

$$\pi = g\sigma_t \text{PB} = g\sigma_0 (1 - \theta t) \text{PB}, \qquad (2)$$

by means of which the atmospheric pressure at any place can be calculated.

53. Dalton's and Gay-Lussac's Law of the Expansion of Gases by Heat.—The following experimental law was discovered by Gay-Lussac † and Dalton, and more recently corrected by Regnault.

If the pressure remains constant, an increase of temperature of 1° C. produces in a given mass of air an expansion of .003665 of its volume.

By means of this experimental law, combined with Boyle's (Art. 48), the relation between the pressure, density, and temperature of a given mass of air or gas may be expressed.

Conceive that a mass of air at the temperature of 0° C. is inclosed in a cylinder by a piston to which a given force is

---

* Enc. Brit., Vol. XVI., p. 33.   † See Deschanel's Nat. Phil., p. 207.

applied; let the temperature be increased to t; the piston will then be forced out until the original volume v_0 is increased by $.003665 t v_0$, where v_0 is the volume of air at 0°.

Let v be the volume of the same mass of air at the temperature t; then we have

$$v = v_0 (1 + .003665 t);$$

or, denoting .003665 by α, we have

$$v = v_0 (1 + \alpha t). \tag{1}$$

Cor. 1.—If Fahrenheit's scale is used, the number of degrees above the freezing point is $t - 32$; and, since 180° F. correspond to 100° C., the expansion for 1° F. is $\dfrac{.3665}{180} = \dfrac{1}{492}$ of the volume at 32° F. The more accurate value of the denominator is 491.13.

Hence, the increase of volume $= \dfrac{v_0 (t - 32)}{492}$;

and, for the whole volume, we have

$$v = v_0 + \dfrac{v_0 (t - 32)}{492},$$

or,
$$v = v_0 \dfrac{460 + t}{492}, \tag{2}$$

where t is the temperature on Fahrenheit's scale, and v_0 is the volume at 32° F.

Cor. 2.—If v' be the volume which the same mass of air assumes at the temperature t', we have

$$v' = v_0 \dfrac{460 + t'}{492}. \tag{3}$$

Dividing (3) by (2), we have

$$v' = v \dfrac{460 + t'}{460 + t}. \tag{4}$$

By means of (4) we may determine the volume which a gas will assume at a given temperature; or, conversely, the temperature it will have under a given volume, if the volume it has at any given temperature is known, the pressure remaining constant.

EXAMPLES.

1. If 100 cubic inches of gas at 68° F. be heated to 120° F., find the volume, the pressure being constant.

Ans. 109.85 cu. ins.

2. A mass of air at 50° F. is raised to 51° F. What is the increase of its volume under a constant pressure?

Ans. $\frac{1}{510}$ of its volume.

54. Law of the Pressure, Temperature, and Density of a Mass of Gas.—Let p, ρ, and v be the pressure, density, and volume of a mass of gas at the temperature t, v_0 and ρ_0 the volume and density at 0.

Then, when p remains constant, we have, from (1) of Art. 53,
$$v = v_0 (1 + at). \qquad (1)$$

Now, if t remains constant while the gas is compressed from v to v_0, the volume varies inversely as the density (Boyle's Law); that is,
$$v : v_0 :: \rho_0 : \rho,$$

which in (1) gives,
$$\rho_0 = \rho (1 + at). \qquad (2)$$

Substituting in (2) of Art. 48, we have
$$p = k\rho_0 = k\rho (1 + at). \qquad (3)$$

COR. 1.—If p', ρ' be the pressure and density of the same gas at a temperature t', we have
$$p' = k\rho' (1 + at');$$

$$\therefore \frac{p}{p'} = \frac{\rho}{\rho'} \frac{1+at}{1+at'}. \tag{4}$$

Cor. 2.—If the volume, and therefore the density, remains constant, while the temperature rises, the pressure will also rise.

Let p_0 be the pressure when $t = 0$, v and ρ remaining constant. Then (3) becomes,

$$p_0 = k\rho. \tag{5}$$

Substituting in (3), we have

$$p = p_0(1 + at), \tag{6}$$

where p and p_0 are the pressures at the temperatures t and 0, the volume being constant.

Let $t = 1$, then (6) becomes

$$p - p_0 = p_0 a = .003665 p_0 \text{ (Art. 53)};$$

that is, *if the volume of a mass of gas remains constant, an increase of temperature of* 1° C. *produces an increase of pressure equal to* .003665 *of its original pressure.*

Cor. 3.—If Fahrenheit's scale is used, (3), (4), and (6) become respectively

$$p = k\rho \frac{460 + t}{492}, \tag{7}$$

$$\frac{p}{p'} = \frac{\rho}{\rho'} \frac{460 + t}{460 + t'}, \tag{8}$$

$$p = p_0 \frac{460 + t}{492}. \tag{9}$$

Cor. 4.—If p' be the pressure of the same gas at a temperature t', the volume remaining constant, we have, from (9),

$$p' = p_0 \frac{460 + t'}{492};$$

$$\therefore \quad \frac{p}{p'} = \frac{460 + t}{460 + t'}, \tag{10}$$

in which p and p' are the pressures corresponding to the temperatures t and t' of a given mass of gas, the volume being constant.

COR. 5.—Since the volume of a given mass of air varies inversely as its density, we have, from (4) and (8),

$$v' = v \frac{1 + at'}{1 + at} \cdot \frac{p}{p'}, \tag{11}$$

$$v' = v \frac{460 + t'}{460 + t} \cdot \frac{p}{p'}, \tag{12}$$

where v' and v denote the volumes of a given mass of air at the temperatures t' and t.

EXAMPLES.

1. If the pressure of a given mass of gas be 29.25 inches, at the temperature 56° F., what will it become if heated to 300° F., the volume being constant?   *Ans.* 43.081 inches.

2. If 200 cubic inches of gas at 60° F., under a pressure of 30 inches of mercury, be raised in temperature to 280° F., while the pressure is reduced to 20 inches, find the volume.
Ans. 426.9 cubic inches.

55. Absolute Temperature.—If we can imagine the temperature of a gas lowered until its pressure vanishes, without any change of volume, we arrive at what is called the absolute zero of temperature, and *absolute temperature* is measured from this point.*

Let t_0 represent this temperature on the Centigrade scale; then (3) of Art. 54 becomes

---
* Besant's Hydromechanics, p. 113.

$$0 = k\rho(1 + \alpha t_0), \quad (1)$$

or,
$$t_0 = -\frac{1}{\alpha} = -273°.$$

In Fahrenheit's scale, the reading for absolute zero is $-459°$.

Combining (1) of this Art. with (3) of Art. 54, we have

$$p = k\rho\alpha(t - t_0)$$
$$= k\rho\alpha(t + 273) = k\rho\alpha T, \quad (2)$$

where T is the absolute temperature.

If v and ρ be the volume and density of a mass of gas, ρv is constant, and therefore, from (2), $\frac{pv}{T}$ is constant; from which it appears that *the product of the pressure and volume of a given mass of gas is proportional to the absolute temperature.*

SCH.—If the difference of temperature between the freezing and boiling points be divided into a hundred degrees, as in the Centigrade thermometer, the freezing point will then be 273° and the boiling point 373° absolute temperature, and the zero of the scale will be that temperature at which the pressure vanishes. Denoting the absolute temperature by T, and the ordinary Centigrade temperature by t, we have
$$T = 273° + t. \quad (3)$$

56. The Pressure of a Mixture of Gases.—If two *liquids*, which do not act chemically on each other, are mixed together in a vessel which remains at rest, they will gradually separate, and finally attain equilibrium with the lighter liquid above the heavier. But if two *gases* are placed in communication with each other, even if the heavier be below the lighter, they will rapidly intermingle until the proportion of the two gases is the same throughout, and the greater the difference of density the more rapidly will the mixture take place.

Take two different gases, of the same temperature and pressure, contained in separate vessels; let a communication be established between the vessels, and it will be found that, unless a chemical action take place, the two gases will permeate each other till they are completely mixed, and that, when equilibrium is attained, the pressure of the mixture will be the same as before, provided the temperature is the same. Hence, from this experimental fact, the following proposition can be deduced.

57. Mixture of Equal Volumes of Gases having Unequal Pressures.—*If two gases having the same temperature be mixed together in a vessel of volume v, and if the pressures of the gases when respectively contained in v, at the same temperature, be p and p', the pressure of the mixture will be $p+p'$.*

Suppose the gases are separate. Take the gas whose pressure is p, and change its volume until its pressure is p', its temperature remaining the same. Its volume will then be, by Mariotte's law (Art. 48), $\dfrac{p}{p'} v$.

Now let the two gases be mixed without change of volume, so that the volume of the mixture is

$$v + \frac{p}{p'} v = \frac{p + p'}{p'} v;$$

then the pressure of the mixture will be p', according to the preceding experimental fact (Art. 56). Now if the mixture be compressed till its volume is v, its temperature remaining constant, the pressure will become, by Mariotte's law, $p + p'$.

This result is equally true for a mixture of any number of gases.

58. Mixture of Unequal Volumes of Gases having Unequal Pressures.—*Two volumes v, v', of different gases, at the respective pressures p, p', are mixed together so that the volume of the mixture is V; to find the pressure of the mixture.*

Change the volume of each gas to V; their pressures will be, respectively (Art. 48),

$$\frac{v}{V}p, \qquad \frac{v'}{V}p',$$

and therefore (Art. 57) the pressure of the mixture is

$$\frac{v}{V}p + \frac{v'}{V}p';$$

and if P be this pressure, we have

$$PV = pv + p'v'.$$

(See Besant's Hydromechanics, p. 114.)

59. Vapors, Gases.—The term *vapor* is applied to those gaseous bodies, such as steam, which can be liquefied at ordinary pressures and temperatures; while the word *gas* generally denotes a body which, under ordinary conditions, is never found in any state but the gaseous. The laws already stated of gases are equally true of vapors within certain ranges of temperature, the only difference between the *mechanical* qualities of vapors and gases, as distinguished from their *chemical* qualities, being that the former are easily condensed into liquids by lowering the temperature, while the latter can be condensed only by the application either of great pressure or extreme cold, or a combination of both.

Prof. Faraday succeeded in condensing a number of different gases; he found that carbonic acid, at the temperature of $-11°$, was liquefied by a pressure of 20 atmospheres, but when it was at the temperature of 0°, a pressure of 36 atmospheres * was required to produce condensation.

In 1877, M. Pictet succeeded in liquefying oxygen by subjecting it to a pressure of 300 atmospheres; at the close of the same year, M. Cailletet effected the liquefaction of nitrogen, hydrogen, and atmospheric air. Such experimental results point to the general conclusion that all gases are the vapors of liquids of different kinds.†

60. Formation of Vapor, Saturation.—The majority of liquids, when left to themselves in contact with the atmosphere, gradually pass into the state of vapor and disappear. This phenomenon occurs much more rapidly with some liquids than with others. Thus, a drop of ether disappears almost instantaneously; alcohol also evaporates very quickly; but water evaporates much more slowly. If water be introduced into a space containing dry air, vapor is immediately formed; if the temperature be increased, or the space enlarged, the quantity of vapor will be increased; but if the temperature be lowered, or the space diminished, some portion of the vapor will be condensed; in all cases the pressure of the air will be increased by the pressure due to the vapor thus formed. The formation of vapor is independent of the presence of air or of its density, the only effect which the air produces being a retardation of the time in which the vapor is formed. If water be introduced into a vacuum, it is instantaneously filled with vapor, but the quantity of vapor is the same as if the space had been originally filled with air.

While the supply of water remains, as a source from which vapor can be produced, any given space will be always *saturated* with vapor, *i. e.*, there will be as much vapor as the temperature admits of. If the temperature be lowered, a portion of the vapor will be immediately condensed, and become visible in the form of a liquid; but if

---

* An atmosphere denotes the pressure due to a column of mercury 29.9 inches in height.

† Besant's Hydrostatics, p. 136.

the temperature be increased so that all the water is turned into vapor, then for this and all higher temperatures the pressure of the vapor will change in accordance with the same law which regulates the connection between the pressure and temperature of gases (Art. 53).

The atmosphere always contains more or less aqueous vapor, and if p be the pressure of dry air, and π of the vapor in the atmosphere at any time, the actual pressure of the atmosphere is $p + \pi$.

61. Volume of Atmospheric Air without its Vapor.—*Having given the pressures of a volume v of atmospheric air, and of the vapor it contains, to find the volume of the air without its vapor at the same pressure and temperature.*

Let P be the pressure of the atmosphere and p that of the vapor; and let v' be the required volume of the air without its vapor, at the pressure P. Then $P - p$ is the pressure of the air alone when its volume is v. Hence we have (Art. 48),

$$P : P - p :: v : v';$$

$$\therefore \ v' = \frac{P - p}{P} v.$$

62. Pressure of Gas when Volume and Temperature are Changed.—*A gas contained in a closed vessel of volume v is in contact with water, and its pressure at the temperature t is P; it is required to determine its pressure when v is changed to v' and t to t'.*

Let p and p' be the pressures of the vapor at the temperatures t and t', respectively, and P' the required pressure.

Then $P - p$ and $P' - p'$ are the pressures of the gas alone, under the two sets of conditions stated. Hence, calling ρ and ρ' the densities of the gas, we have, from (3) of Art. 54,

$$P - p = k\rho(1 + \alpha t),$$
$$P' - p' = k\rho'(1 + \alpha t');$$

also, from (1) of Art. 48, we have $v\rho = v'\rho'$.

$$\therefore \frac{P' - p'}{P - p} = \frac{v}{v'} \frac{1 + \alpha t'}{1 + \alpha t}, \qquad (1)$$

which gives the value of P'.

Cor.—If σ and σ' be the densities of vapor under the two conditions, we have

$$\frac{p'}{p} = \frac{\sigma'(1 + \alpha t')}{\sigma(1 + \alpha t)}. \qquad (2)$$

Dividing (1) by (2), we get

$$\frac{p}{p'} \frac{P' - p'}{P - p} = \frac{v\sigma}{v'\sigma'};$$

or, $$\frac{v'\sigma'}{v\sigma} = \frac{Pp' - pp'}{P'p - pp'}. \qquad (3)$$

If $Pp' > P'p$, $v'\sigma'$ will exceed $v\sigma$; i. e., more vapor will have been absorbed by the gas. But if $Pp' < P'p$, then $v'\sigma'$ will be less than $v\sigma$, and the gas must therefore, in changing its volume and temperature, have lost a portion of its vapor. (See Besant's Hydrostatics, p. 138.)

EXAMPLE.

Having given the pressures P and p of a volume v of atmospheric air, and of the vapor it contains, to find the volume of the air, without its vapor, at the same pressure P, the temperature remaining constant.

Ans. Volume of air $= \dfrac{P - p}{P} v$.

63. Formation of Dew, the Dew Point.—*Dew* is the name given to those drops of water which are seen in the morning on the leaves of plants, and are especially noticeable in the spring and autumn. If any portion of the space occupied by the atmosphere be saturated with vapor, *i. e.*, if the density of the vapor be as great as it can be for the temperature, then the slightest fall of temperature will produce condensation of some portion of the vapor; but if the density of the vapor be not at its maximum for that temperature, no condensation will take place until the temperature is lowered below the point corresponding to the saturation of the space.

If any body in contact with the atmosphere be cooled down until its temperature is below that which corresponds to the saturation of the air around it, condensation of the vapor will take place, and the condensed vapor will be deposited in the form of *dew* upon the surface of the body. Heat radiates from the ground, and from the bodies upon it, and unless there are clouds from which the heat would be radiated back, the surfaces are cooled, and the vapor in the adjacent stratum of the atmosphere condenses and falls in small drops of water on the surface. The formation of dew on the ground depends therefore on the cooling of its surface, and this is in general greater and more quickly effected when the sky is free from clouds. This accounts for the dew with which the ground is covered after a clear night. A covering of any kind will diminish the formation of dew beneath; for instance, but very little dew will be formed under the shade of large trees.

The *dew-point* is the temperature at which vapor begins to be deposited in the form of dew, and it must be determined by actual observation.

64. Pressure of Vapor in the Air.—Tables * have

---

* Besant's Hydrostatics, p. 143.

been formed and empirical formulæ constructed for determining the relation between the temperature and the elastic force of vapor, at the saturating density, for certain ranges of temperature. If, therefore, the dew-point be ascertained, we can at once determine the pressure of the vapor in the air by means of these tables. For, if t' be the dew-point, and p' the corresponding pressure, then at any other temperature t of the air above t', we have, for the required pressure,

$$p = \frac{1 + at}{1 + at'} p'.$$

65. Effect of Compression or Dilatation on the Temperature of a Gas.—It is an experimental fact that, if a quantity of air be suddenly compressed, its temperature is raised; and that, if the compression be of small amount, the relative increase of temperature is proportional to the condensation. Thus, if the density be changed from ρ to ρ', the increase of temperature is proportional to

$$\frac{\rho' - \rho}{\rho}.$$

If the air be allowed to dilate, its temperature is diminished according to the same law. A stream of compressed air when issuing from a closed vessel is sensibly chilled. The reason that the compression or dilatation must be sudden, is that no heat should be allowed to escape, or to be admitted. If the experiment be performed in a non-conducting vessel, there is no necessity for rapidity of action.

66. Expansion of Bodies — Maximum Density of Water.—In general, all solid and liquid bodies expand under the action of heat, and contract when heat is withdrawn. The expansion of mercury is proportional to the increase of temperature, within certain limits; this is also the case with solid bodies, such as glass and steel. For water and aqueous bodies generally, the law of expansion is unknown.

It is a remarkable property of water that, at a temperature of about 4° C. or 40° F., its volume is a minimum and therefore its density is a maximum;* and whether its temperature increases or decreases from this point, the water expands in volume. When the temperature descends to the freezing point, there is a still further expansion at the moment of congelation; for this reason, ice floats in water.

We can now see what takes place in a pond of fresh water during winter. The fall of temperature at the surface of the pond does not extend to the bottom, where the water seldom falls below 4° C., whatever may be the external temperature. As the temperature at the surface descends, the water at the surface cools, and being contracted, it becomes heavier than the water beneath, and sinks to the bottom. The water from beneath rises and becomes cooled in its turn; and this process goes on till all the water has attained its maximum density, *i.e.*, till its temperature is 4° C. But when all the water has attained this temperature, it will remain stationary; and any further cooling of the water at the surface will expand it, until it finally congeals. It is clear that the deeper the water is, the longer will be the time before the whole of the water has attained its maximum density, and therefore that ice will form much less rapidly on the surface of deep than on the surface of shallow ponds.

It is from the fact that water expands in freezing, taken in connection with the low conducting power of liquids generally, that the temperature at the bottom of deep ponds remains moderate even during very severe cold, and that the lives of aquatic animals are preserved.

---

* The results of Playfair and Joule give 3°.945 C. as the temperature at which the density is a maximum. Phil. Transactions, 1856.

67. Thermal Capacity — Unit of Heat — Specific Heat.

The *thermal capacity* of a body is the quantity of heat required to raise the temperature of the body one degree.

The *unit of heat* which is generally employed is the quantity of heat required to raise a unit mass of water through one degree C., the temperature of the water being between 0° C. and 40° C. It is called the thermal unit Centigrade.

The *specific heat* of a body is the thermal capacity of a unit of its mass; and it is always to be understood that the same unit of mass is employed for the body as for the water mentioned in the definition of the unit of heat. Therefore, specific heat is independent of the unit, and is merely the ratio of the quantity of heat required to increase by 1° the temperature of the body to the quantity of heat required to increase by 1° the temperature of an equal mass of water.

The quantity of heat expended in changing the temperature from t to t'

varies as $t' - t$ when the mass is given,

and varies as the mass when $t' - t$ is given;

and therefore generally it varies as $m(t' - t)$, if m be the mass. Hence, the quantity of heat expended in changing the temperature of the mass m from t to t' is

$$sm(t' - t), \qquad (1)$$

where s is the specific heat of the substance, since it is the quantity of heat required to raise by 1° the temperature of the unit of mass, which may be shown by putting $m = 1$ and $t' - t = 1$.

Let dH denote the quantity of heat which produces in the unit of mass a change of temperature dt, then the measure of the specific heat is $\frac{dH}{dt}$.

68. Comparison of Specific Heat at a Constant Pressure with that at a Constant Volume.

—In the specific heat of gases there are two cases to be considered: (1) when the pressure remains constant, the gas being allowed to expand; (2) when the volume is constant.

Let the pressure p remain constant while the application of a small quantity of heat H increases the temperature T by τ, and changes the density from ρ to ρ'. From (2) of Art. 55, by putting $k\alpha = K$, we have

$$p = K\rho T = K\rho'(T + \tau). \qquad (1)$$

Now if the air be rapidly compressed into its original volume, its temperature will be increased (Art. 65), and we shall have

$$\frac{\text{the increase of temperature}}{T} = \mu \frac{\rho - \rho'}{\rho'}$$

$$= \mu \frac{\tau}{T} \text{ [by (1)]},$$

where μ is a constant.

$$\therefore \text{ the increase of temperature} = \mu\tau, \qquad (2)$$

and hence the whole change of temperature produced by the heat H, when the volume is constant,

$$= \tau + \mu\tau = \lambda\tau. \qquad (3)$$

In order, therefore, to produce a change of temperature τ when the volume is constant, the quantity of heat required is $\dfrac{H}{\lambda}$, and consequently,

$$\frac{\text{specific heat at constant pressure}}{\text{specific heat at constant volume}} = \frac{H}{\dfrac{H}{\lambda}} = \lambda. \qquad (4)$$

Cor.—Therefore the specific heat at a constant pressure exceeds the specific heat at a constant volume; and this excess from (2) is equal to the quantity of heat $\mu\tau$ that is disengaged when the gas is suddenly compressed into its original volume.

SCH.—The value of λ is found experimentally to be constant for all simple gases, its value being approximately 1.408. (See Besant's Hydromechanics, p. 118.)

EXAMPLES.

1. A mass m_1 of a substance of specific heat s_1 and temperature t_1, is mixed with a mass m_2 of a substance of specific heat s_2 and temperature t_2, the mixture being merely mechanical, so that no heat is generated or absorbed by any action between the substances, and all gain or loss of heat from external sources is prevented. Find the resulting temperature t of the mixture.

Suppose the former body to be the warmer; then it cools down from t_1 to t, while the colder rises from t_2 to t. Therefore we shall have

$$m_1 s_1 (t_1 - t) = \text{the units of heat lost by the former body,}$$

and $$m_2 s_2 (t - t_2) = \text{the units of heat gained by the latter body,}$$

and since the quantity of heat lost by the warmer body is equal to that gained by the cooler, these two expressions are equal; therefore

$$m_1 s_1 (t_1 - t) = m_2 s_2 (t - t_2);$$

$$\therefore t = \frac{m_1 s_1 t_1 + m_2 s_2 t_2}{m_1 s_1 + m_2 s_2}.$$

One of the methods of finding the specific heat of a substance is by immersing it in a given weight of water, and observing the temperature attained by the two substances.

2. A mass M of a substance of specific heat S and temperature T, is immersed in a vessel of water, m' and m being the masses of the vessel and of the water in it, and t' their common temperature and s' the specific heat of the vessel. Find the temperature t of the whole after immersion.
$$\text{Ans. } t = \frac{MST + mt' + m's't'}{MS + m + m's'}.$$

3. A glass vessel weighing 1 lb. contains 5 oz. of water, both at 20°, and 2 oz. of iron at 100° is immersed. What is the temperature of the whole, taking .2 as the specific heat of glass and .12 of iron? Ans. $22°\tfrac{58}{211}$.

The following are approximate values of the specific heats of a few substances:

Water,	1
Thermometer-glass,	0.198
Iron,	0.114
Zinc,	0.1
Mercury,	0.03
Silver,	0.06
Brass,	0.09

(Besant's Hydrostatics, p. 147.)

69. Sudden Compression of a Mass of Air.—*A mass of air being suddenly* compressed or dilated, it is required to find the new pressure and temperature.*

Let p, ρ, T be the pressure, density, and absolute temperature at any stage of the process; p', ρ', T' the new pressure, density, and temperature; and let dT be the change of temperature due to the change $d\rho$ in ρ. Then we have

$$\frac{dT}{T} = \mu \frac{d\rho}{\rho}. \qquad (1)$$

---

* If the compression takes place in a non-conducting vessel, so that no heat is lost or gained, the compression need not be rapid.

From (1) of Art. 68, we have

$$p = K\rho T; \qquad (2)$$

$$\therefore \frac{dp}{d\rho} = KT + K\rho \frac{dT}{d\rho}$$

$$= KT + K\mu T \text{ [from (1)]}. \qquad (3)$$

Dividing (3) by (2), we have

$$\frac{1}{p}\frac{dp}{d\rho} = \frac{1}{\rho} + \frac{\mu}{\rho} = \frac{\lambda}{\rho} \text{ [from (3) of Art. 68]},$$

or,
$$\frac{dp}{p} = \frac{\lambda\, d\rho}{\rho}. \qquad (4)$$

Integrating between the limits p' and p, ρ' and ρ, we have

$$\frac{p'}{p} = \left(\frac{\rho'}{\rho}\right)^{\lambda};$$

$$\therefore p' = p\left(\frac{\rho'}{\rho}\right)^{\lambda}, \qquad (5)$$

which determines the pressure.

Also, $\qquad p' = K\rho' T',$

which, divided by (2), gives

$$\frac{p'}{p} = \frac{\rho' T'}{\rho T} \qquad (6)$$

From (5) and (6), we have

$$\frac{T'}{T} = \left(\frac{\rho'}{\rho}\right)^{\lambda-1};$$

$$\therefore T' = T\left(\frac{\rho'}{\rho}\right)^{\lambda-1}, \qquad (7)$$

which determines the temperature. (See Besant's Hydromechanics, p. 118.)

70. Mass of the Earth's Atmosphere.—By means of the barometer, some idea may be formed of the mass of air and vapor surrounding the earth, since the weight of the whole atmosphere is equal to that of a stratum of mercury about 29.9 inches thick covering the globe. Suppose the earth to be a sphere of radius r, and that h is the height of the barometric column at all points of its surface. Then the mass of the atmosphere is approximately equivalent to the mass $4\pi\sigma r^2 h$ of mercury, where σ is the density of the mercury.

Let ρ be the mean density of the earth; then,

the mass of the atmosphere : the mass of the earth

$$= 4\pi\sigma r^2 h : \tfrac{4}{3}\pi\rho r^3$$

$$= 3\sigma h : \rho r.$$

Taking $\sigma = 13.568$ (Art. 47), and $\rho = 5.5$,* and supposing the height of the barometric column h to be 30 inches, which is probably near the average height at sea-level,† it will be found that the above ratio of the mass of the atmosphere to that of the earth is about $\frac{1}{1141600}$.

71. The Height of the Homogeneous Atmosphere.—If the atmosphere were of the same density throughout as at the surface of the earth, its height l would be approximately obtained from the following equation,

$$\sigma h = \rho l, \qquad (1)$$

where σ and ρ are the densities of mercury and air respectively, and h is the height of the barometric column. From Art. 70, and Art. 33, Sch., we have

$$\sigma = 13.568 \times 768\rho = 10420.224\rho,$$

---

* There is some doubt about the accuracy of this value; the value deduced by the Astronomer Royal at the Harton Colliery in 1854 is 6.6. Phil. Trans., 1856.

† See Ency. Brit., Vol. III., p. 28.

and taking $h = 30$ inches, we have, by solving (1) for l,

$$l = h\frac{\sigma}{\rho} = 26050 \text{ feet,}$$

which is a little less than 5 miles.

72. Necessary Limit to the Height of the Atmosphere.—Since the attraction of the earth diminishes at a distance from its surface (Anal. Mechs., Art. 133a), it is clear that the atmosphere is very far from being of uniform density throughout, and therefore the result in Art. 71 is very far from the truth. A *limit* can be found, however, to the height of the atmosphere from the consideration that, beyond a certain distance from the earth's centre, its attraction will be unable to retain the particles of air in the circular paths which they describe about the earth, since the centrifugal force must exceed the force of gravity.

Let ω be the earth's angular velocity, and r its radius. Then the centrifugal force of a particle m of air on the earth's surface is $m\omega^2 r$, and this is equal to $\dfrac{mg}{289}$ [Anal. Mechs., Art. 199, (3)]; therefore, at a height z above the surface, the centrifugal force $m\omega^2(r+z)$

$$= \frac{mg}{289}\frac{r+z}{r}.$$

The earth's attraction at the same height (Anal. Mechs., Art. 133a)

$$= \frac{mgr^2}{(r+z)^2};$$

and, in order that the particle may be retained in its path, these two forces must equal each other.

$$\therefore \frac{mg}{289}\frac{r+z}{r} = \frac{mgr^2}{(r+z)^2},$$

or, $$\frac{r+z}{289r} = \frac{r^2}{(r+z)^2},$$

$$\therefore z = r(\sqrt[3]{289} - 1);$$
$$= 5.6r +$$
$$= 22000 \text{ miles (approximately).}$$

Rem.—The actual height of the atmosphere, however, is possibly much lower than this, for its temperature has been found, by experiments made in balloons, to diminish with great rapidity during an ascent; it is therefore very likely that, at a height less than $5r$, the air may be liquefied by extreme cold, and in that case its external surface would be of the same kind as the surfaces of known inelastic fluids. (Besant's Hydromechanics, p. 120.)

73. Decrease of Density of the Atmosphere.—
(1) *When the force of gravity is constant.*

Take a vertical column of the atmosphere, and let it be divided into an indefinite number of horizontal strata of equal thickness, so that the density of the air may be uniform throughout the same stratum. Let the weight of the whole column from the top of the atmosphere to the earth $= a$, that of the whole column above the lowest stratum $= b$, that of the column above the second $= c$, and so on. Then b, c, d, etc., are the forces respectively which compress the first, second, third, etc. strata, which, as they are of equal thickness, are as their weights, $a - b$, $b - c$, $c - d$, etc. Hence we have

$$a - b : b - c :: b : c;$$
$$\therefore a : b :: b : c.$$

In the same way, it may be shown that

$$b : c :: c : d,$$

and so on. Hence, b, c, d, etc., and therefore the *densities* of the successive strata, form a series of terms in geometric progression, which is decreasing since a is greater than b, and therefore b greater than c, and so on; and as the strata all have the same thickness, the heights of the several strata above the earth's surface increase in arithmetic progression. Hence,

If a series of heights be taken in arithmetic progression, when the force of gravity is constant, the densities of the air decrease in geometric progression.

SCH.—By barometric observations at different altitudes, it is found that at the height of $3\frac{1}{2}$ miles above the earth's surface, the air is about one-half as dense as it is at the surface. Forming therefore an arithmetic series, with $3\frac{1}{2}$ for the common difference, to denote the heights, and a geometric series with $\frac{1}{2}$ for the common ratio, to denote densities, we have

Heights, $3\frac{1}{2}$, 7, $10\frac{1}{2}$, 14, $17\frac{1}{2}$, 21, $24\frac{1}{2}$, 28, $31\frac{1}{2}$, 35, etc.
Densities, $\frac{1}{2}$, $\frac{1}{4}$, $\frac{1}{8}$, $\frac{1}{16}$, $\frac{1}{32}$, $\frac{1}{64}$, $\frac{1}{128}$, $\frac{1}{256}$, $\frac{1}{512}$, $\frac{1}{1024}$, etc.

That is, according to this law, at the height of 35 miles the air is less than a thousandth part as dense as it is at the surface of the earth.

(2) *When the force of gravity varies inversely as the square of the distance from the earth's centre.*

Let r be the radius of the earth, ρ' the density at the surface of the earth, ρ the density at a height z, and h the height of a homogeneous atmosphere. Then, since the density varies as the compressing force, and this varies as the weight, we have

$$\rho' : d\rho :: h\rho' g : g\frac{r^2}{(r+z)^2}(-\rho\, dz),$$

where g and $\dfrac{gr^2}{(r+z)^2}$ are the measures of the earth's attraction at the surface and at a height z, the negative sign being taken because the density is a decreasing function of the height z.

$$\therefore \frac{d\rho}{\rho} = -\frac{r^2}{h} \frac{dz}{(r+z)^2}.$$

Integrating, observing that when $z = 0$, $\rho = \rho'$, we have

$$\log \frac{\rho}{\rho'} = \frac{r^2}{h}\left(\frac{1}{r+z} - \frac{1}{r}\right);$$

$$\therefore \rho = \frac{\rho'}{e^{\frac{r^2}{h}\left(\frac{1}{r} - \frac{1}{r+z}\right)}},$$

which shows that, if $r + z$ increases in harmonic progression, $\dfrac{1}{r+z}$ will decrease in arithmetic progression, and therefore ρ will decrease in geometric progression. Hence,

If a series of heights be taken in harmonic progression, when the force of gravity is regarded as variable, the densities of the air decrease in geometric progression. (See Bland's Hydrostatics, p. 258.)

74. Heights Determined by the Barometer.—A very important use of the barometer is to find the difference of level of two places situated at unequal distances above the surface of the earth. Since the height of the column of mercury in the barometer depends on the pressure of the atmosphere (Art. 43), and as the pressure of the atmosphere at any point depends upon the height of the column of air extending from that point to the top of the atmosphere, it follows that this pressure will decrease as we ascend above the earth's surface, and therefore that the height of the column of mercury will diminish. That is, the mercury in the barometer will fall when the instrument is carried from the foot to the top of a mountain, and will rise again when it is returned to its former position.

(1) *When the force of gravity is regarded as constant.*

Consider a vertical column of the atmosphere at rest under the action of gravity. Let z be taken vertical and positive upwards; and at a height z, let p be the pressure and ρ the density. The pressure p, at any height z, is measured by the weight of the column of air extending from that height to the top of the atmosphere; and the elementary pressure dp will be measured by the weight of the column having the same base and the elementary height dz. Therefore, if A be the area of the section of the column, we have

$$A dp = - A g\rho \, dz,$$

or,
$$dp = - g\rho \, dz, \qquad (1)$$

the negative sign being taken because the pressure p is a decreasing function of the height z.

If t be the temperature, we have from (3) of Art. 54,

$$p = k\rho (1 + \alpha t). \qquad (2)$$

Dividing (1) by (2), we have

$$k \frac{dp}{p} = - \frac{g \, dz}{1 + \alpha t}. \qquad (3)$$

If the heights above the earth's surface are small, the force of gravity g may be regarded as constant; and supposing t constant, we have, by integrating (3),

$$k \log \frac{p}{p'} = \frac{g(z' - z)}{1 + \alpha t}, \qquad (4)$$

where p' is the pressure at the height z'.

Let h, h' be the observed barometric heights at the two stations, whose altitudes are z and z'; let σ be the density of mercury at a temperature zero, and τ, τ', the temperatures at the two stations. Then we have, from (2) of Art. 52,

$$p = g\sigma h(1 - \theta\tau),$$

and
$$p' = g\sigma h'(1 - \theta\tau'),$$

which in (4) gives

$$z' - z = \frac{k}{g}(1 + \alpha t)\log\frac{h(1-\theta\tau)}{h'(1-\theta\tau')}, \quad (5)$$

where t may be taken approximately equal to $\frac{1}{2}(\tau + \tau')$; from this equation the difference of the heights of the two stations can be calculated.

(2) *When the force of gravity is regarded as variable.*

If the heights above the earth's surface be considerable, it is necessary to take account of the variation of gravity at different distances from the earth's centre.

Calling g the measure of the earth's attraction at the level of the sea, and r the radius of the earth, then we have, for the measure of the attraction at a height z,

$$g\frac{r^2}{(r+z)^2}, \quad (6)$$

which, being substituted in (1) for g, gives

$$dp = -g\frac{r^2}{(r+z)^2}\rho\, dz. \quad (7)$$

Dividing (7) by (2), we have

$$k\frac{dp}{p} = -\frac{1}{1+\alpha t}\frac{gr^2\, dz}{(r+z)^2}. \quad (8)$$

It must be observed that p is the sum of the pressures due to the air itself, and to the aqueous vapor which is mixed with it; *i. e.*, the quantity kp in (2) is the sum of the two, $k\rho$, $k'\rho'$, where ρ and ρ' are the densities of the air and the aqueous vapor, respectively.

Considering t constant as before, and equal to the mean of the temperatures at the two stations, and integrating (8), we have

$$k \log \frac{p'}{p} = \frac{gr^2(z-z')}{(1+\alpha t)(r+z)(r+z')}. \tag{9}$$

As before, let h, h', and τ, τ', be the observed barometric heights and temperatures, and σ the density of mercury at a temperature zero; then from (2) of Art. 52, by substituting for g its value from (6), we have

$$p = \frac{gr^2}{(r+z)^2} \sigma h (1 - \theta\tau),$$

$$p' = \frac{gr^2}{(r+z')^2} \sigma h' (1 - \theta\tau'),$$

$$\therefore \frac{p'}{p} = \left(\frac{r+z}{r+z'}\right)^2 \frac{1-\theta\tau'}{1-\theta\tau} \frac{h'}{h}. \tag{10}$$

Substituting (10) in (9), and solving for $z - z'$, we have

$$z - z' =$$
$$\frac{k(1+\alpha t)(r+z)(r+z')}{gr^2}\left(\log\frac{h'}{h} + 2\log\frac{r+z}{r+z'} + \log\frac{1-\theta\tau'}{1-\theta\tau}\right). \tag{11}$$

Since θ is very small (Art. 52), we have

$$\log\frac{1-\theta\tau'}{1-\theta\tau} = \log[1 - \theta(\tau' - \tau)]$$

$$= -\theta(\tau' - \tau).$$

(Calculus, Art. 61.)

Substituting this in (11), and reducing Naperian to common logarithms by multiplying by m, the modulus of the common system, we have

$$z - z' = \frac{k(1+\alpha t)(r+z)(r+z')}{mgr^2}\left[\log_{10}\frac{h'}{h} + 2\log_{10}\frac{r+z}{r+z'} - m\theta(\tau' - \tau)\right], \quad (12)$$

from which the value of z can be determined when z' is known.

COR. 1.—If the lower station be nearly at the level of the sea, $z' = 0$, and (12) becomes

$$z = \frac{k(1+\alpha t)}{mg}\left(1+\frac{z}{r}\right)\left[\log_{10}\frac{h'}{h} + 2\log_{10}\left(1+\frac{z}{r}\right) - m\theta(\tau'-\tau)\right]. \quad (13)$$

COR. 2.—In the above investigation no account has been taken of the variation of gravity at different parts of the earth's surface. From a comparison of the results obtained by causing pendulums to oscillate in different latitudes, if g be the measure of gravity at a place of latitude λ, and g' at a place of latitude λ', it has been found (Poisson, Art. 628) that

$$\frac{g}{g'} = \frac{1 - .002588 \cos 2\lambda}{1 - .002588 \cos 2\lambda'};$$

therefore,
$$\frac{k}{mg} = \frac{k}{mg'}\cdot\frac{1-.002588\cos 2\lambda'}{1-.002588\cos 2\lambda}. \quad (14)$$

If λ' be the latitude of Paris, the value of the quantity

$$\frac{k}{mg'}(1 - .002588 \cos 2\lambda') \quad (15)$$

is nearly 18336 French metres,* or about 60158.56 English

---

* A French metre is 39.37079 inches.

feet; representing this numerical quantity by c and substituting it in (14), we get

$$\frac{k}{mg} = \frac{c}{1 - .002588 \cos 2\lambda},$$

which in (13) gives

$$z = \frac{c(1+\alpha t)\left(1+\frac{z}{r}\right)}{1-.002588 \cos 2\lambda}\left[\log_{10}\frac{h'}{h} + 2\log_{10}\left(1+\frac{z}{r}\right) - m\theta(\tau'-\tau)\right], \quad (16)$$

from which the value of z may be determined by a series of approximations; *i. e.*, an approximate value must be first obtained by neglecting $\frac{z}{r}$; then this approximate value must be substituted for z in $\frac{z}{r}$, and a more accurate value will be obtained, and the same process may be repeated, if necessary.*

Sch. 1.—When $\frac{z}{r}$ is very small, it may be neglected in (16). It has been found in practice, however, that in this case the results are more accurate by employing 18,393 metres as the value of c. (Duhamel, p. 259.)

In order that the heights as determined by the barometer may be very exact in practice, certain corrections are necessary. For instance, the value of k is modified by the fact that the density of aqueous vapor at a given temperature and pressure is less than the density of dry air under the same circumstances; and the proportion of aqueous vapor to dry air will generally be different at the two stations.

---

* A formula for this is given in Ency. Brit., Vol. III., p. 386, involving a consideration of densities of vapor.

Sch. 2.—Formula (16) has been obtained on the supposition that the temperature of the air remains constant in passing from the lower to the higher station; if, however, the difference between the heights be very great, a considerable error may be thus introduced, and formulæ have therefore been constructed in which account is taken, on various hypotheses, of the variation of atmospheric temperature. A formula of this kind is given in Lindeman's Barometric Tables, constructed on the supposition that the temperature diminishes in harmonic progression through a series of heights increasing in arithmetic progression.

Also, we have assumed that the temperature of the mercury in the barometer is the same as that of the air surrounding it; but in some cases, as for instance when observations are made in a balloon, the barometer may not remain long enough in the same place to acquire the temperature of the surrounding air. The temperature of the mercury may be observed, however, by placing the bulb of a thermometer in the cistern of the barometer, and the temperatures thus obtained must be used in (10). (See Besant's Hydromechanics, p. 121.)

SPECIFIC GRAVITIES.

Ratios of the Specific Gravities of different substances to that of water at 60°.

Diamond,	3.52	Tin,	7.29
Sulphur,	2	Lead,	11.45
Iodine,	4.94	Zinc,	6.86
Arsenic,	5.96	Nickel,	8.38
Gold,	19.4	Iron,	7.844
Platina,	21.53	Flint-glass,	2.5
Silver,	10.5	Marble,	2.716
Mercury,	13.568	Rock-salt,	1.92
Copper,	8.85	Ivory,	1.917

EXAMPLES. 131

Ice (at 0°), . . . 0.926 | Alcohol, . . . 0.794
Sea-water, . . . 1.027 | Ether, . . . 0.724
Olive-oil, . . . 0.915

Ratios of the densities of gases and vapors of different substances to that of atmospheric air at the same temperature and under the same pressure.

Oxygen, ... 1.103 | Water, ... 0.62
Hydrogen, . . . 0.069 | Alcohol, ... 1.613
Nitrogen, . . . 0.976 | Carbonic Acid, . 1.524
Chlorine, . . . 2.44 | Ammonia, . . . 0.591
Bromine, . . . 5.395 | Sulphurous Acid, 2.212
Iodine, ... 8.701 | Sulphuric Acid, . 2.763
Arsenic, ... 10.365 | Ether, ... 2.586
Mercury, . . . 6.978

EXAMPLES.

1. If the barometer stand at 28.372 inches, find the pressure on a square inch. *Ans.* 13.902 lbs.

2. If the elastic force of a vapor sustain a column of mercury 3.34 inches high, find its pressure on a square inch. *Ans.* 1.64 lbs.

3. A cubic inch of mercury at 16° weighs 3429½ grs. nearly, and the barometer stands at 30 inches. Find (1) the atmospheric pressure on the square inch of surface, and (2) the height of a barometer filled with water instead of mercury, the specific gravity of mercury being 13.6.
Ans. (1) 14.698 lbs.; (2) 34 feet.

4. A hollow cylinder, open at the top, is inverted, and partly immersed in water. It is required to find the depth of the surface of the water within the cylinder below the surface of the water without.

Let a = the length of the cylinder, b = the length of the part not immersed, x = the required depth of the surface within below the surface without, and π, π', the pressures of the atmospheric air and of the compressed air. Then (Art. 48) we have

$$\pi' : \pi :: a : b + x; \qquad (1)$$

also, $\pi' =$ pressure on the water within $= \pi + g\rho x = g\rho h + g\rho x$, if h be the height of the water barometer.

Substituting these values of π and π' in (1), we have

$$\frac{h+x}{h} = \frac{a}{b+x};$$

$$\therefore \quad x = \frac{\sqrt{4ah + (h-b)^2} - (h+b)}{2}$$

5. A cylinder, 20 ft. long, is half filled with water, and inverted with the open end just dipping into a vessel of water. Find the altitude of the water in the cylinder, the height of the water barometer being 33 feet.
Ans. 7.21 feet.

6. When the mercurial barometer stands at 30 inches, what is the height of the barometer formed of a liquid whose specific gravity is 5.6 ? *Ans.* 72.7 inches nearly.

7. The air contained in a cubical vessel, the edge of which is one foot, is compressed into a cubical vessel of which the edge is one inch. Compare the pressures on a side of each vessel. *Ans.* 1 : 12.

8. If the elastic force of a mass of gas whose volume is 100 cubic inches be 30.275 inches of mercury, find its elastic force if it be allowed to expand to a volume of 387 cubic inches. *Ans.* 7.823 inches.

9. If Fahrenheit's Thermometer mark 40°, what are the corresponding marks of Reaumur's and the Centigrade ?
Ans. $4\frac{4}{9}$; $3\frac{5}{9}$.

EXAMPLES. 133

10. If the sum of the readings on Fahrenheit's and the Centigrade thermometer be zero for the same temperature, find the reading of each thermometer.

Ans. $11\frac{3}{4}$; $-11\frac{3}{4}$.

11. If 327 cubic inches of gas at 280° be allowed to cool down to 56°, find the volume.* *Ans.* 223 cubic inches.

12. If, by the application of heat, 120 cubic inches at 60° F. expand into 180 cubic inches, find the temperature.

Ans. 320°.

13. If the pressure be 14.7 lbs. on the square inch at the temperature 62°, what will it become if raised to 420°?

Ans. 24.78 lbs.

14. If the pressure at 50° be 15 lbs., and if the temperature be so far increased as to make the pressure 21 lbs., find the temperature. *Ans.* 254°.

15. The air in a spherical globe, one foot in diameter, is compressed into another globe, 6 inches in diameter, and the temperature is raised $t°$. Compare (1) the pressures of the air under the two conditions, and (2) the pressures on the surfaces of the globes. *Ans.* $\begin{cases} (1)\ 1 : 8\,(1 + at)\,; \\ (2)\ 1 : 2\,(1 + at). \end{cases}$

16. The temperature of the air in an extensible spherical envelope is gradually raised t', and the envelope is allowed to expand till its radius is n times its original length. Compare the pressure of the air in the two cases.

Ans. $1 + at : n^3$.

17. A mass of air at a temperature t is contained in a cylinder which has an air-tight piston fitting into it, and it is found that the air exerts a pressure P on the piston; the air being suddenly compressed into $\dfrac{1}{n}$ of its former vol-

---

* Fahrenheit's Thermometer is understood, unless otherwise **expressed.**

ume, and the temperature changed to t', find the pressure P' on the piston.

$$Ans.\ P' = Pn\frac{1+at'}{1+at}.$$

18. If a cubic foot of gas, whose temperature is 100° and elastic force 29½ inches, be cooled down to 40°, and compressed by a force equivalent to 10½ inches, find its volume.
Ans. 4334.7 cubic inches.

19. If 20 cubic inches of air, whose temperature is 56° and elastic force 28.8 inches, be expanded to 25 inches by the application of heat, and if the elastic force become 31 inches, find the temperature. Ans. 234.27°.

20. Let 100 cubic inches of air have a temperature 32° and a pressure 29.922 inches; if the temperature become 60°, and the pressure 30 inches, find the volume.
Ans. 105.42 cubic inches.

21. A cubic foot of air at a temperature of 100°, and under a pressure of 29½ inches of mercury, is cooled down to 40° and compressed by an additional 10½ inches of mercury. Find the volume. Ans. 1137.86 cubic inches.

22. If h and h' be the heights of the surface of the mercury in the tube of a barometer above the surface of mercury in the cistern at two different times, compare the densities of the air at those times, the temperature being supposed unaltered. Ans. $h : h'$.

23. A conical wine-glass is immersed, mouth downwards, in water. How far must it be depressed in order that the water within the glass may rise half-way up it?
Ans. $7h$, where h is the height of the water barometer.

24. A cubic foot of air having a pressure of 15 lbs. on a square inch is mixed with a cubic inch of compressed air, having a pressure of 60 lbs. on a square inch. Find the pressure of the mixture when its volume is 1729 cubic inches. Ans. $15\tfrac{45}{1729}$ lbs.

25. Two volumes, V and V', of different gases, at pressures p, p', and temperature t, are mixed together; the volume of the mixture is U, and its temperature t'. Determine the pressure.

Ans. $\dfrac{pV + p'V'}{U} \cdot \dfrac{1 + at'}{1 + at}$.

26. Three gallons of water at 45° are mixed with six gallons at 90°. What is the temperature of the mixture?

Ans. 75°.

27. An ounce of iron at 120°, and 2 oz. of zinc at 90°, are thrown into 6 oz. of water at 10°, contained in a glass vessel weighing 10 oz. What is the final temperature, taking .1 and .12 as the specific heats of zinc and iron?

Ans. $13°\tfrac{53}{104}$.

PART II.
HYDROKINETICS.

CHAPTER I.

MOTION OF LIQUIDS.—EFFLUX.—RESISTANCE AND WORK OF LIQUIDS.

75. Velocity of a Liquid in Pipes.—*If a liquid run through any pipe of variable diameter, which is kept continually full, and the velocity is the same in every part of a transverse section, the velocities in the different transverse sections vary inversely as the areas of the sections.*

For as the tube is kept full, and the liquid is incompressible (Art. 3), the same quantity of liquid which runs through one section will, in the same time, run through the next section, and so on through any other. Hence if k, k' be the areas of any two sections, and v, v' the velocities of the particles at those sections, we have, since the quantity of liquid which flows through any section in a unit of time is the product of the area of the section by the velocity,

$$kv = k'v';$$

$$\therefore \; v : v' :: k' : k. \qquad (1)$$

Cor.—Hence, as the section of a mass of liquid decreases, its velocity increases in the same proportion. For instance, the velocity of a stream or river is greater at places where its width is diminished. This demonstration is also applicable to different sections of a liquid issuing through the orifice of a vessel, whether the section be taken within or without the vessel, provided there be no vacuity in the stream between the sections.

Sch.—It is supposed in this proposition that the changes in the diameters of the sections are *gradual*, and nowhere abrupt; if there are any angles in the pipe, they will produce eddies in the motion of the liquid, and the proposition will not hold true.

76. Velocity of Efflux.—*If a small aperture be made in a vessel containing liquid, the velocity with which the liquid issues from the vessel is the same as if it had fallen from the level of the surface to the level of the aperture.**

Let EF represent a very small orifice in the bottom of the vessel ABCD, which is filled with a liquid to the level AB; and suppose the vessel to be kept full by supplying it from above, while the liquid is running out through the orifice EF. Let v be the velocity of efflux, w the weight of the liquid which issues with that velocity per second, and h the height of the surface above the orifice, called the *head*† of the liquid. Then the work which w can perform while descending through the distance h, from the surface to the orifice $= wh$, and the kinetic energy stored up in w as it issues through the orifice $= \dfrac{w}{2g} v^2$ (Anal. Mechs., Art. 217). If we suppose there is no loss of energy during the passage through the orifice,

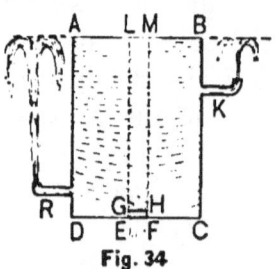

Fig. 34

---

* This is known as Torricelli's Theorem.

† The term *head* in Hydromechanics is measured, relatively to any point, by the depth of that point below the surface of the liquid. Since the liquid in Fig. 34 descends through a height h to the orifice, we may say there are h feet of head above the orifice.

we may equate these two quantities of work, and shall have
$$\frac{wv^2}{2g} = wh,$$
from which we find
$$v^2 = 2gh; \tag{1}$$
$$\therefore\ v = \sqrt{2gh}; \tag{2}$$

that is, *the velocity of efflux is the same as that of a body which has fallen freely through the height h.*

From (2) we have $h = \dfrac{v^2}{2g}$, in which the height h, corresponding to the velocity v, is called the *head due to the velocity*, or simply *the head*. The corresponding velocity is called *the velocity due to the head*.

Cor. 1.—If the orifice be made in the vertical face of the vessel, and a tube be inserted so as to direct the current *obliquely, horizontally,* or *vertically upward,* the velocity of efflux will be the same, since the pressure of fluids at the same depth is the same in every direction (Art. 7), and each particle of liquid having the same velocity will follow the same path; a parabola whose directrix, whatever be the angle of elevation, is fixed, and lies in the surface of the liquid (Anal. Mechs., Arts. 151 and 153). If the liquid issue obliquely, its equation is given in (3) of Art. 151, Anal. Mechs. If it issue horizontally, $\alpha = 0$, and this equation becomes
$$x^2 = \frac{2v^2}{g} y = 4hy.$$

Cor. 2.—If h_1 be the depth of a second orifice below the surface, and v_1 the velocity, we have
$$v_1 = \sqrt{2gh_1}; \tag{3}$$
therefore, from (2) and (3), we have
$$v : v_1 :: \sqrt{h} : \sqrt{h_1};$$

VELOCITY OF EFFLUX.

that is, *the velocities of efflux are as the square roots of the depths.*

COR. 3.—The quantity of liquid run out in any time is equal to a cylinder, or prism, whose base is the area of the orifice, and whose altitude is the space described in that time by the velocity acquired in falling through the height of the liquid.

COR. 4.—If any pressure be exerted on the surface of the liquid, the velocity of efflux will be increased.

Let h be the depth of the orifice below the surface of the liquid, h_1 the height of the column of liquid which would exert the same pressure as that which is applied at the surface; then the velocity of efflux will be due to the vertical height $h + h_1$; hence we have from (2)

$$v = \sqrt{2g(h + h_1)}. \qquad (4)$$

If h_1 be taken equal to the height of a column of water equal to the pressure of the atmosphere ($= 34$ feet), (4) becomes

$$v = \sqrt{2g(h + 34)}. \qquad (5)$$

which is the velocity of efflux when a liquid is projected into a vacuum, the orifice being at a depth h, below the surface of the liquid.

If k be the area of the orifice, then the quantity of liquid Q which flows through the orifice in the unit of time is

$$Q = kv = k\sqrt{2gh}. \qquad (6)$$

COR. 5.—If a parabola, with a parameter $= 2g$, be described with its axis vertical, and vertex in the upper surface of the liquid, the velocity of efflux through any small orifices in the side, would be represented by the corresponding ordinates.

Sch.—The correctness of this theorem can also be shown by the following experiment. If in the vessel (Fig. 34) an orifice K or R be made, directed vertically upwards, the velocity of the jet K or R is such as to carry the particles of liquid up nearly to the same level as the surface of the liquid in the vessel. Practically the resistance of the air and friction in the conducting tube destroy a portion of this velocity.

EXAMPLES.

1. With what velocity will water issue from a small orifice $16\frac{1}{12}$ ft. below the surface of the liquid?

Ans. $32\frac{1}{6}$ ft.

2. A vessel has in it a hole an inch square; water is kept in the basin at a constant level of 9 ft. above the hole; what is the outflow in one hour?   *Ans.* 600 cu. ft.

3. What is the discharge per second through an orifice of 10 square inches, 5 ft. below the surface of the liquid?

Ans. 2152 cu. ins.

77. The Horizontal Range of a Liquid Issuing through a very Small Orifice in the Vertical Side of a Vessel.—Let ABCD be a vessel filled with a liquid, having its side BC vertical, M a small orifice in the side of the vessel, MH the parabola described by the liquid, and CH the horizontal range. On BC describe the semicircle BFC, and through M draw MN perpendicular to BC. If the liquid issue horizontally from the orifice M, the equation of its path is (Art. 76, Cor. 1),

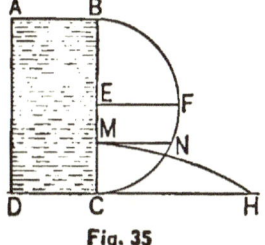

Fig. 35

$$x^2 = 4hy, \qquad (1)$$

in which $h = $ BM, the height of the surface above the

orifice; then the range CH will be determined by making $y = MC$. Hence we have from (1)

$$x = 2\sqrt{hy} = 2\sqrt{BM \times MC}$$
$$= 2MN ; \qquad (2)$$

that is, *the horizontal range of a liquid issuing horizontally through a very small orifice in the side of a vessel is equal to twice the ordinate at the orifice, in a semicircle whose diameter is the vertical distance from the surface of the liquid to the horizontal plane.*

COR.—When the orifice is made at the centre of the side BC, the horizontal range is a maximum, and equal to the height of the liquid above CH; at equal distances above and below the centre, the range will be the same.

78. Time of Discharge from a Cylindrical Vessel when the Height is Constant.—*When a cylindrical vessel is kept constantly full, it is required to determine the time in which a quantity of liquid equal in volume to the cylinder will flow through a small orifice in its base.*

Let h be the height of the surface, K the area of the base of the vessel, and k of the orifice, V the velocity of descent of the surface of the liquid, and v the velocity of efflux at the orifice, and t the time necessary to discharge a volume of liquid equal to that of the cylinder, which remains constantly full.

Then the quantity of liquid which flows through the orifice in the unit of time is $k\sqrt{2gh}$; and since the velocity of the surface is V, the quantity of liquid which passes through the orifice in the unit of time must equal VK. Hence we have

$$VK = k\sqrt{2gh};$$

$$\therefore V = \frac{k}{K}\sqrt{2gh}\,; \qquad (a)$$

and as the vessel is kept constantly full, we have

$$t = \frac{h}{V} = \frac{hK}{k\sqrt{2gh}} = \frac{Q}{k\sqrt{2gh}}, \qquad (1)$$

where Q denotes the whole quantity of liquid in the vessel.

Cor.—If the liquid be kept at a height h' in a second vessel, containing a quantity Q', which flows through an orifice k', in the time t, we have from (1)

$$t = \frac{Q'}{k'\sqrt{2gh'}}\,; \qquad (2)$$

and from (1) and (2) we have

$$Q : Q' :: k\sqrt{h} : k'\sqrt{h'}.$$

Hence, *the quantities discharged in the same time, from orifices of different sizes, and at different depths, are as the areas of those orifices and the square roots of their depths jointly.*

79. The Time of Emptying any Vessel through a Small Orifice in the Bottom.—Let EH be the upper surface of the liquid at the time t, x and y the distances OD and DH, h the depth OC of the liquid when the vessel is full, k the area of the orifice, and K the area of the upper surface of the liquid at the time t, which, when the figure of the vessel is known, will be given in terms of x and y.

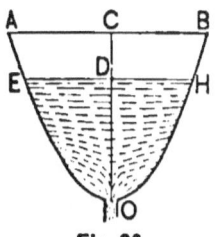

Fig. 36

Then the quantity of liquid which flows through the orifice in an element of time is $k\sqrt{2gx}\,dt$; and since in the same time the surface EH descends a distance dx, the quantity of liquid which flows through the orifice in this time must equal Kdx. Hence we have

$$k\sqrt{2gx}\,dt = -Kdx,$$

the negative sign being taken, because x decreases as t increases,

$$\therefore\ t = -\int \frac{Kdx}{k\sqrt{2gx}}. \tag{1}$$

Cor. 1.—If the vessel be a surface of revolution round a vertical axis, $K = \pi y^2$, which in (1) gives

$$t = -\pi \int \frac{y^2 dx}{k\sqrt{2gx}}. \tag{2}$$

Cor. 2.—To determine the time of emptying a right cylinder or prism. Here K is constant, and (1) becomes

$$t = -\frac{K}{k\sqrt{2g}}\int \frac{dx}{\sqrt{x}} = -\frac{2K}{k\sqrt{2g}}x^{\frac{1}{2}} + C$$

$$= \frac{2K}{k\sqrt{2g}}(\sqrt{h} - \sqrt{x}), \tag{3}$$

remembering that when $t = 0$, $x = h$.

When $x = 0$, we have for the time of emptying the whole cylinder,

$$t = \frac{2K}{k\sqrt{2g}}\sqrt{h} = \frac{2Q}{k\sqrt{2gh}}, \tag{4}$$

where Q denotes the quantity of liquid in the vessel.

By comparing this result with that in (1) of Art. 78, it appears that *the time necessary for the entire discharge of the liquid when the vessel empties itself is twice as great as that which is required to discharge the same quantity when the vessel is kept constantly full.*

Cor. 3.—If a cylinder of given altitude empty itself in

n seconds, through a given orifice, the radius r of the cylinder from (4) is

$$r = \sqrt{\frac{nk\sqrt{g}}{\pi\sqrt{2h}}}, \qquad (5)$$

and if the radius is given, its height h is

$$h = \frac{n^2 k^2 g}{2\pi^2 r^4}. \qquad (6)$$

80. The Time of Emptying a Cylinder into a Vacuum.—*To determine the time in which a cylindrical vessel will empty itself, through an orifice in the bottom, into a vacuum, when its upper surface is exposed to the pressure of the atmosphere.*

Let h be the height of the vessel, h' the height of a column of liquid which is equal to the weight of the atmosphere; and x the depth of the orifice below the upper surface of the liquid. Then from (4) of Art. 76, the velocity of discharge is $\sqrt{2g(x+h')}$, which in (1) of Art. 79 gives

$$t = -\frac{K}{k\sqrt{2g}} \int \frac{dx}{\sqrt{x+h'}}$$

$$= -\frac{2K}{k\sqrt{2g}}(x+h')^{\frac{1}{2}} + C$$

$$= \frac{2K}{k\sqrt{2g}}\left[(h+h')^{\frac{1}{2}} - (x+h')^{\frac{1}{2}}\right], \qquad (1)$$

since when $x = h$, $t = 0$.

And making $x = 0$, (1) becomes

$$t = \frac{2K}{k\sqrt{2g}}\left[(h+h')^{\frac{1}{2}} - h'^{\frac{1}{2}}\right], \qquad (2)$$

which is the time of emptying the vessel.

81. The Time of Emptying a Paraboloid.—Let the vessel be a paraboloid of revolution round the vertical axis, h its height, and $2p$ its parameter. Then if x is the depth of the orifice in the bottom below the upper surface of the liquid, we have

$$y^2 = 2px,$$

which, in (2) of Art. 79, gives

$$t = -\frac{2p\pi}{k\sqrt{2g}} \int \frac{xdx}{\sqrt{x}} = -\frac{4p\pi}{3k\sqrt{2g}} x^{\frac{3}{2}} + C$$

$$= \frac{4}{3}\frac{p\pi}{k\sqrt{2g}} (h^{\frac{3}{2}} - x^{\frac{3}{2}}), \tag{1}$$

since when $x = h$, $t = 0$.

Making $x = 0$, and putting $r = $ the radius of the base, (1) becomes

$$t = \frac{2}{3}\frac{r^2\pi h^{\frac{1}{2}}}{k\sqrt{2g}}, \tag{2}$$

which is the time of emptying the vessel.

82. Cylindrical Vessel with Two Small Orifices.—A cylindrical vessel of given dimensions, is filled with a liquid; there are two given and equal small orifices, one at the bottom, the other bisecting the altitude; to find the time of emptying the upper half, supposing both orifices to be opened at the same instant.

Let $2a = $ the altitude of the vessel, $x = $ the altitude of the surface of the liquid from the upper orifice at the time t, and $r = $ the radius of the base. Then the quantities of liquid which flow through the upper and lower orifices in one second are, respectively, $k\sqrt{2gx}$ and $k\sqrt{2g(x+a)}$, which in (1) of Art. 79, gives

$$t = -\frac{\pi r^2}{k\sqrt{2g}} \int \frac{dx}{\sqrt{x} + \sqrt{x+a}}$$

$$= -\frac{\pi r^2}{k\sqrt{2g}} \int \frac{\sqrt{a+x} - \sqrt{x}}{a} dx$$

$$= \frac{2\pi r^2}{3ka\sqrt{2g}} \left[(2\sqrt{2} - 1) a^{\frac{3}{2}} - (a+x)^{\frac{3}{2}} + x^{\frac{3}{2}} \right], \quad (1)$$

between the limits, $x = a$ and $x = x$.

And making $x = 0$, (1) becomes

$$t = \frac{4\pi r^2}{3k} \sqrt{\frac{a}{2g}} (\sqrt{2} - 1), \quad (2)$$

which is the time of emptying the upper half of the vessel. (See Bland's Hydrostatics, p. 165.)

83. Orifice in the Side of a Conical Vessel.—*A hollow cone, base downward, whose vertical angle is 60°, is filled with a liquid; to determine the place where a small orifice must be made in its side, so that the issuing liquid may strike the horizontal plane in a point whose distance from the bottom of the vessel is to the distance of the orifice from the top* $:: 5 : 4$.

Let $AN = x$, and $AM = a$; then $NM = a-x$, and $AP = \dfrac{2x}{\sqrt{3}}$.

Also by hypothesis we have

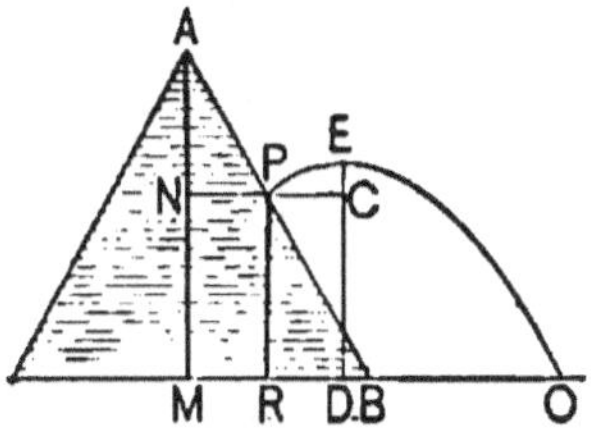

$$BO : AP :: 5 : 4;$$

$$\therefore BO = \frac{5x}{2\sqrt{3}}.$$

$$BR = PR \tan 30° = \frac{a-x}{\sqrt{3}};$$

$$\therefore RO = \frac{a-x}{\sqrt{3}} + \frac{5x}{2\sqrt{3}} = \frac{2a + 3x}{2\sqrt{3}}.$$

Fig. 37

Substituting this value of RO for x in (3) of Art. 151, Anal. Mechs., and for v^2 its value $2gx$, we have

$$x - a = \frac{2a + 3x}{2\sqrt{3}} \tan 30° - \left(\frac{2a + 3x}{2\sqrt{3}}\right)^2 \frac{1}{4x \cos^2 30°}$$

$$= \frac{2a + 3x}{6} - \frac{(2a + 3x)^2}{36x};$$

$$\therefore \quad x = \tfrac{2}{3}a(1 \pm \sqrt{\tfrac{2}{3}});$$

which is the depth of the orifice from the vertex. (See Bland's Hydrostatics, p. 142.)

84. Velocity of Efflux through an Orifice of any Size in the Bottom of a Cylindrical Vessel.—Let AB be the upper surface of the liquid at a height h above the orifice EF; consider any lamina GH, at a height x above the orifice; and as before, let k, K be the areas of the orifice and the section GH, respectively.

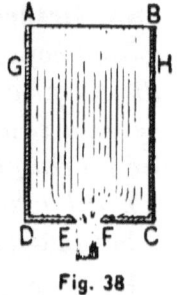

Fig. 38

At the height x above the orifice, let p be the pressure and ρ the density, and at a height $x + dx$, let $p + dp$ be the pressure. Then the volume Kdx of liquid may be considered as acted upon by the pressures pK, $(p + dp)K$, and its weight, $-g\rho Kdx$. Hence the moving force will be

$$pK - (p + dp)K - g\rho Kdx = -Kdp - g\rho Kdx;^*$$

and since the moving force is measured by the mass into the acceleration (Anal. Mechs., (3) of Art. 20), we have

$$-Kdp - g\rho Kdx = \rho Kdx \frac{d^2x}{dt^2}.$$

$$\therefore \quad \frac{d^2x}{dt^2} = -\frac{dp + g\rho dx}{\rho dx}. \tag{1}$$

---

* The moving force is here negative because x is positive upwards.

Let v be the velocity of efflux; then the quantity of liquid which flows through the orifice in an element of time is $kvdt$, and since in the same time the surface K descends a distance equal to dx, we have

$$Kdx = -kvdt, \quad \text{or} \quad \frac{dx}{dt} = -\frac{kv}{K}, \qquad (2)$$

the negative sign being taken because x decreases as t increases;

$$\therefore \frac{d^2x}{dt^2} = -\frac{k}{K}\frac{dv}{dt},$$

which in (1) gives

$$\frac{k}{K}\frac{dv}{dt} = \frac{dp + g\rho dx}{\rho dx}$$

or,

$$dp + g\rho dx = \frac{\rho k dv}{K}\frac{dx}{dt} = -\frac{\rho k^2}{K^2} v dv \quad \text{[from (2)]}.$$

Integrating, we have

$$p + g\rho x = -\frac{\rho k^2 v^2}{2K^2} + C$$
$$= p + \frac{\rho v^2}{2}\left(1 - \frac{k^2}{K^2}\right),$$

remembering that when $x = 0$, $K = k$.

Hence,
$$2gx = v^2\left(1 - \frac{k^2}{K^2}\right);$$

$$\therefore v = \sqrt{\frac{2gx}{1 - \frac{k^2}{K^2}}}; \qquad (3)$$

which is the velocity of efflux at a depth x.

When $x = h$, or the vessel is full, we have for the velocity

$$v = \sqrt{\frac{2gh}{1 - \frac{k^2}{K^2}}}. \qquad (4)$$

Cor. 1.—As the ratio $\dfrac{k}{K}$ of the sections decreases the velocity decreases, becoming a minimum and $= \sqrt{2gh}$, when the cross-section k of the orifice is very small compared with that of K, which agrees with (2) of Art. 76, as it clearly should.

Cor. 2.—As the ratio $\dfrac{k}{K}$ increases the velocity increases, and it approaches nearer and nearer to infinity, the smaller the difference between the two cross-sections becomes. If $k = K$, (4) becomes

$$v = \frac{\sqrt{2gh}}{0} = \infty;$$

from which we infer that, if a cylindrical vessel is without a bottom, a liquid must flow in and out with an infinitely great velocity, or else a section of the liquid flowing out of the vessel can never be equal to a section of the vessel. If a cylindrical tube be vertical, and filled with a liquid, the portion of the liquid at the lower extremity, being urged by the pressure of all above it, will necessarily have a greater velocity than those portions which are higher, and therefore (Art. 75) a section of the liquid issuing from the vessel must be less than a section of the tube, *i.e.*, the stream of liquid will not fill the orifice of exit.*

EXAMPLE.

If water flows from a vessel, whose cross-section is 60 square inches, through a circular orifice in the bottom 5 inches in diameter, under a head of water of 6 feet, find its velocity. *Ans.* 20.79 ft.

85. Rectangular Orifice in the Side of a Vessel.—*To determine the quantity of liquid which will*

---

* Formula (4) was first given by Bernouilli, and was afterwards much disputed (Weisbach's Mechanics, p. 804).

flow from a rectangular orifice in the side of a vessel which is kept constantly full.

(1) *When one side of the orifice coincides with the surface of the liquid.*

Let h be the height and b the breadth of the rectangular orifice ALMD, through which the efflux takes place; let HK be a horizontal strip at the distance x below AD, and of infinitesimal thickness dx, so that the velocity of the liquid in every part of the strip is the same.

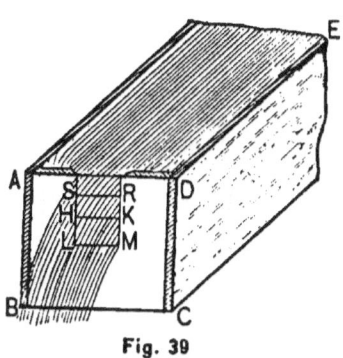

Fig. 39

Then the velocity of efflux through this strip is $\sqrt{2gx}$ [Art. 76, (2)], and the quantity discharged in a unit of time is $b\,dx\sqrt{2gx}$; hence, calling Q the whole quantity discharged in a unit of time, we have

$$Q = \int b\,dx\,\sqrt{2gx}\,; \qquad (1)$$

and integrating between $x = 0$ and $x = h$, we have

$$Q = \tfrac{2}{3} b\,\sqrt{2gh^3}. \qquad (2)$$

If we denote by v the *mean* velocity, i.e., the velocity which would have to exist at every point of the orifice, in order that the same quantity of liquid would flow through the orifice with a uniform velocity as now flows through with the variable velocity, we have

$$Q = bhv,$$

which in (2) gives

$$v = \tfrac{2}{3}\sqrt{2gh}, \qquad (3)$$

Hence *the mean velocity of a liquid flowing out through a rectangular orifice in the side of a vessel* is ⅔ the velocity at the lower edge of the orifice; and the quantity of liquid flowing out through this orifice in any given time is ⅔ the quantity that would flow **through** an orifice of equal area placed horizontally at the *whole* depth, in the same time, the vessel being kept constantly full.

(2) *When the upper surface of the rectangular orifice is below the surface of the liquid.*

Let SR be the upper edge of the orifice at the depth h_1 below the surface AD. Then, integrating (1) between the limits $x = h_1$ and $x = h$, we have

$$Q = \tfrac{2}{3} b \sqrt{2g} \left(h^{\frac{3}{2}} - h_1^{\frac{3}{2}} \right). \qquad (4)$$

If the *mean* velocity of efflux is v, we have

$$Q = b(h - h_1)v,$$

which in (4) gives

$$v = \tfrac{2}{3} \sqrt{2g} \, \frac{h^{\frac{3}{2}} - h_1^{\frac{3}{2}}}{h - h_1}. \qquad (5)$$

86. Triangular Orifice in the Side of a Vessel.—

(1) *When the vertex of the triangle is in the surface of the liquid.*

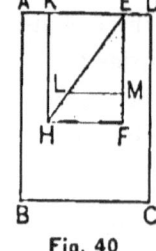

Fig. 40

Let h be the height EF, and b the breadth HF of the triangular orifice EHF, through which the efflux takes place; let LM be a horizontal strip at the distance x below AD, and of infinitesimal thickness dx, so that the velocity of the liquid in every part of the strip is the same.

Then $LM = \dfrac{b}{h} x$, and calling Q the quantity of liquid discharged in a unit of time, we have

$$Q = \int_0^h \frac{b}{h} x \sqrt{2gx}\, dx$$

$$= \tfrac{2}{5} b \sqrt{2gh^3}. \qquad (1)$$

If the mean velocity is v, we have

$$Q = \tfrac{1}{2} bhv,$$

which in (1) gives $\qquad v = \tfrac{4}{5}\sqrt{2gh}. \qquad (2)$

(2) *When the base of the triangle is in the surface of the liquid.*

Let KEH be the triangular orifice, $KE = b$, and $KH = h$. Then the quantity discharged through KEH will equal the discharge through the rectangle KHFE, minus that through the triangle EHF; therefore subtracting (1) of this Art. from (2) of Art. 85, we have

$$Q = \tfrac{2}{3} bh \sqrt{2gh} - \tfrac{2}{5} bh \sqrt{2gh}$$

$$= \tfrac{4}{15} bh \sqrt{2gh}, \qquad (3)$$

and $\qquad v = \tfrac{8}{15}\sqrt{2gh}. \qquad (4)$

Cor. 1.—If the orifice be a trapezoid ABCD, whose upper base $AB = b_1$ lies in the surface of the liquid, whose lower base $CD = b_2$, and whose altitude is $DF = h$, the discharge may be found by combining the discharge through the rectangle ECDF with those through the two triangles ADF and BCE. Hence, combining (3) with (2) of Art. 85, we have

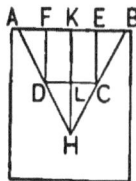

Fig. 41

$$Q = \tfrac{2}{3} b_2 h \sqrt{2gh} + \tfrac{4}{15}(b_1 - b_2) h \sqrt{2gh}$$

$$= \tfrac{2}{15}(2b_1 + 3b_2) h \sqrt{2gh}. \qquad (5)$$

Cor. 2.—If the orifice be a triangle DCH (Fig. 41), whose base $DC = b_1$ is situated at a depth $KL = h_1$ below the surface of the liquid, and whose vertex H is at a depth h below the surface, the discharge is equal to that through the triangle AHB, minus that through the trapezoid ABCD. Hence, from (3) and (5), we have

$$Q = \tfrac{4}{15} bh \sqrt{2gh} - \tfrac{2}{15}(2b + 3b_1) h_1 \sqrt{2gh_1}$$

$$= \tfrac{2}{15}\sqrt{2g}\left[2b(h^{\frac{3}{2}} - h_1^{\frac{3}{2}}) - 3b_1 h_1^{\frac{3}{2}}\right]. \qquad (6)$$

Since $AB : DC :: HK : HL$, we have

$$b : b_1 :: h : h - h_1 ;$$

$$\therefore \ b = \frac{b_1 h}{h - h_1},$$

which in (6) gives

$$Q = \frac{2\sqrt{2g}\, b_1}{15} \left(\frac{2h^{\frac{5}{2}} - 5hh_1^{\frac{3}{2}} + 3h_1^{\frac{5}{2}}}{h - h_1} \right). \qquad (7)$$

Cor. 3.—If the orifice be a triangle ABC, whose vertex A is above its base, and at a depth h_1 below the surface of the liquid, whose base $CB = b_1$ is at a depth h below the surface, the discharge is equal to that through the rectangle ACBK, minus that through the triangle ABK. Hence, from (7) and (4) of Art. 85, we have

Fig. 42

$$Q = \left[\tfrac{2}{3} b_1 (h^{\frac{3}{2}} - h_1^{\frac{3}{2}}) - \tfrac{2}{15} b_1 \left(\frac{2h^{\frac{5}{2}} - 5hh_1^{\frac{3}{2}} + 3h_1^{\frac{5}{2}}}{h - h_1} \right) \right] \sqrt{2g}$$

$$= \frac{2\sqrt{2g}\, b_1}{15} \left(\frac{3h^{\frac{5}{2}} - 5h_1 h^{\frac{3}{2}} + 2h_1^{\frac{5}{2}}}{h - h_1} \right). \qquad (8)$$

Otherwise thus: Let ODC be a vertical orifice, formed by a plane curve, whose vertex is O, at the depth AO below the surface of the liquid.

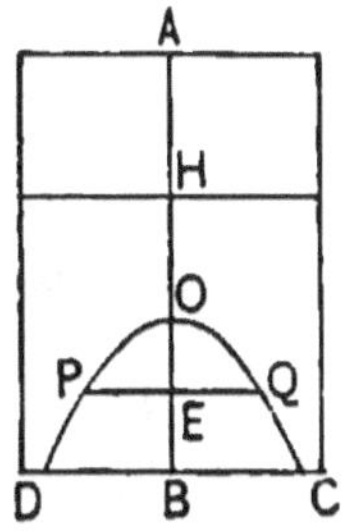

Let $AB = h$, $AO = h_1$, $OE = x$, $EQ = y$; then the area of the horizontal strip PQ, of infinitesimal thickness dx, $= 2y\,dx$; and therefore the quantity discharged in a unit of time through this elemental strip is

$$2y\,dx\,\sqrt{2g(h_1 + x)};$$

and hence we have

$$Q = \int 2y\sqrt{2g(h_1 + x)}\,dx. \tag{9}$$

Fig. 43

(1) *When the orifice is a rectangle.*

Here y is constant, which put $= \tfrac{1}{2}b$, and integrating (9) between the limits $x = 0$ and $x = h - h_1$, we have for the discharge through the whole orifice ODC,

$$Q = \tfrac{2}{3}b\sqrt{2g}\left(h^{\frac{3}{2}} - h_1^{\frac{3}{2}}\right), \tag{10}$$

which is the same as (4) of Art. 85.

Cor. 4.—If the upper side coincides with the surface of the liquid, $h_1 = 0$, and (10) becomes

$$Q = \tfrac{2}{3}bh\sqrt{2gh},$$

which agrees with (2) of Art. 85.

(2) *When the orifice is a triangle whose vertex is downwards and the base horizontal.*

Let $a : b$ be the ratio of the altitude to the base; then

$$2y = \frac{b}{a}(h - h_1 - x),$$

which in (9), and integrating between the limits $x = 0$ and $x = h - h_1$, gives

$$Q = \int_a^b \frac{x}{a}\sqrt{2g}\,(h - h_1 - x)\sqrt{h_1 + x}\,dx$$

$$= \tfrac{2}{15}b\sqrt{2g}\left(\frac{2h^{\frac{5}{2}} - 5hh_1^{\frac{3}{2}} + 3h_1^{\frac{5}{2}}}{h - h_1}\right), \qquad (11)$$

which agrees with (7).

Cor. 5.—If the base coincides with the surface of the liquid, $h_1 = 0$, and (11) becomes

$$Q = \tfrac{4}{15}bh\sqrt{2gh},$$

which agrees with (3).

(3) *When the orifice is a triangle whose vertex is upwards and base horizontal.*

Here $\qquad 2y = \dfrac{b}{a}x,$

which in (9), between the same limits, $x = 0$ and $x = h - h_1$, gives

$$Q = \tfrac{2}{15}b\sqrt{2g}\left(\frac{3h^{\frac{5}{2}} - 5h_1 h^{\frac{3}{2}} + 2h_1^{\frac{5}{2}}}{h - h_1}\right), \qquad (12)$$

which agrees with (8).

Cor. 6.—If the vertex coincides with the surface of the liquid, $h_1 = 0$, and (12) becomes

$$Q = \tfrac{2}{5}bh\sqrt{2gh},$$

which agrees with (1).

Cor. 7.—From Cors. 5 and 6 we see that the quantities discharged in the same time through two equal triangular orifices in the side of a vessel kept constantly full, the one having its base and the other its vertex upwards in the surface of the liquid, are in the ratio of 2 : 3.

87. The Time of Emptying any Vessel through a Vertical Orifice.

Let A be the surface of the liquid in the vessel when the orifice OCD is opened, and H the surface at the end of the time t; let $AH = z$, $AO = h'$, $OE = x$, $AB = h$, and $PQ = 2y$.

Then the quantity discharged through the orifice in an element of time, from (9) of Art. 86, is

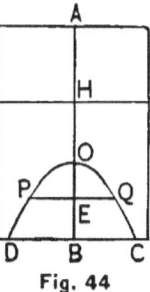

Fig. 44

$$Q = \left[2\sqrt{2g} \int y\sqrt{x+h'-z}\, dx \right] dt, \qquad (1)$$

the x-integration being taken between $h - h'$ and 0, z being constant during this integration; and since, in the same time, the surface of the liquid at H descends a distance dz, the quantity discharged through the orifice in this time must equal $K\, dz$, where K is the area of the section of the vessel at H. Hence, we have

$$\left[2\sqrt{2g} \int y\sqrt{x+h'-z}\, dx \right] dt = K\, dz; \qquad (2)$$

$$\therefore\ t = \frac{1}{2\sqrt{2g}} \int \frac{K\, dz}{\int y\sqrt{x+h'-z}\, dx}, \qquad (3)$$

the z-integration being taken between 0 and h.

EXAMPLE.

Find the time of emptying a cone by an orifice ACB in its side.

Let $AH = h$ be the axis of the cone, $CB = b$, $CA = l$, angle $HAC = \alpha$, $AK = x$, PK being perpendicular to AH. When the orifice is opened, let the surface of the liquid in the vessel be at H, and at the end of the time t let it be at M, and let $AM = z$.

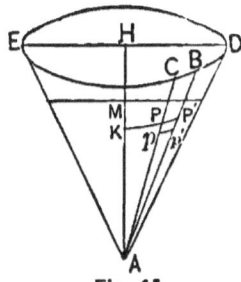

Fig. 45

EXAMPLE.

Then we have

$$AP = x \sec \alpha, \qquad Pp = \sec \alpha \, dx,$$

and
$$y = PP' = \frac{b \sec \alpha}{l} x;$$

$$\therefore \text{ the area of } PP'p'p = \frac{b \sec^2 \alpha}{l} x \, dx.$$

The velocity of discharge through this area

$$= \sqrt{2g(z-x)};$$

therefore the quantity discharged in an element of time

$$= \left[\frac{b \sec^2 \alpha}{l} \sqrt{2g} \int x \sqrt{z-x} \, dx \right] dt$$

$$= \left[\frac{b \sec^2 \alpha}{l} \sqrt{2g} \, \tfrac{4}{15} z^{\frac{5}{2}} \right] dt,$$

the x-limits being 0 and z; and this must equal $K\,dz$, from (2).

Hence we have, from (3), taking the negative sign, because z decreases as t increases,

$$t = \int \frac{-K \, dz}{\frac{b \sec^2 \alpha}{l} \sqrt{2g} \, \tfrac{4}{15} z^{\frac{5}{2}}}$$

$$= \int \frac{-15 l \pi \tan^2 \alpha \, z^2 \, dz}{4b \sqrt{2g} \sec^2 \alpha \, z^{\frac{5}{2}}}$$

$$= -\frac{15 l \pi \tan^2 \alpha}{4b \sqrt{2g} \sec^2 \alpha} \int \frac{dz}{z^{\frac{1}{2}}}$$

$$= \frac{15 \pi l \tan^2 \alpha}{2b \sqrt{2g} \sec^2 \alpha} (\sqrt{h} - \sqrt{z}),$$

between the limits h and z.

Therefore the whole time of emptying the vessel

$$= \frac{15\pi l \tan^2 \alpha \sqrt{h}}{2b\sqrt{2g} \sec^2 \alpha}.$$

(See Bland's Hydrostatics, p. 185.)

88. Efflux from a Vessel in Motion.—If the vessel ABCD be filled with liquid to AB, and raised vertically, with an accelerated motion, by a weight P attached to an inextensible string, without weight, passing over two smooth pulleys F, E, the velocity of efflux is augmented; and if it descends with an accelerated motion, the velocity is diminished.

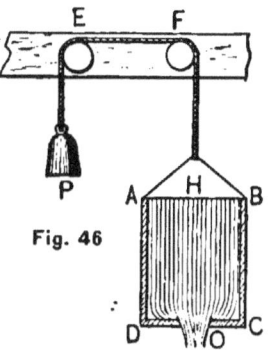

Fig. 46

Let Q be the weight of the vessel and liquid contained in it. Since the pulleys are perfectly smooth, the tension of the string is the same throughout; hence the force which causes the motion is the difference between the weights P and Q. The moving force, therefore, is $P - Q$; but the weight of the mass moved is $P + Q$. Hence, from (1) of Art. 25, Anal. Mechs., we have

$$P - Q = \frac{P + Q}{g} f;$$

$$\therefore f = \frac{P - Q}{P + Q} g, \tag{1}$$

which is the vertical force of acceleration. Since this force acts vertically upwards on the vessel, and the force of gravity g acts vertically downwards, every particle of the liquid presses against the bottom of the vessel, not only with its own weight Mg, but also with its inertia Mf. Hence the entire accelerating force pressing against every point in the base is

$$g + f = g + \frac{P-Q}{P+Q} g \quad \text{[from (1)]}$$

$$= \frac{2P}{P+Q} g. \tag{2}$$

Let $HO = h$, and $v =$ the velocity of efflux; then we have

$$v = \sqrt{2(g+f)h} \tag{3}$$

$$= \sqrt{2g}\sqrt{\frac{2Ph}{P+Q}}. \tag{4}$$

COR. 1.—If the vessel is allowed to empty itself through the orifice O, without receiving any liquid, let $x =$ the variable altitude OH, K the horizontal section of the vessel, which is a function of x, and k the section of the orifice. Then we have

$$Q = \int K \, dx;$$

which in (4) gives for the quantity discharged in an element of time,

$$k\sqrt{2g}\sqrt{\frac{2Px}{P+\int K\,dx}}\,dt = -K\,dx;$$

$$\therefore\; t = \int \frac{-K\,dx\sqrt{P+\int K\,dx}}{k\sqrt{2g}\cdot\sqrt{2Px}}. \tag{5}$$

COR. 2.—If $f = g$, (3) becomes

$$v = \sqrt{2\cdot 2gh} = 2\sqrt{gh};$$

and the velocity of efflux is 1.414 times as great as it would be if the vessel stood still.

Cor. 3.—If in (1), $P = Q$, then $f = 0$, and the vessel is at rest. If $P < Q$, then Q will descend and P ascend, f is negative and (3) becomes

$$v = \sqrt{2(g - f)h},$$

and the vessel descends with an accelerated motion, the velocity being diminished.

Cor. 4.—If $P = 0$, then, from (2), $g + f = 0$, and therefore, from (3), $v = 0$, and there is no pressure on the bottom of the vessel, and no liquid will flow out; which is also evident from this, that every particle in the vessel will descend by its own gravity, with the same velocity.

89. Efflux from a Rotating Vessel.—If a vessel ABCD, containing a liquid, is made to rotate about its vertical axis XX', the surface of the liquid will take the form of a paraboloid of revolution (Art. 21), and at the centre H of the bottom the depth of liquid KH is less than it is near the edge, and the liquid will flow more slowly through an orifice at the centre than through any other orifice of the same size in the bottom.

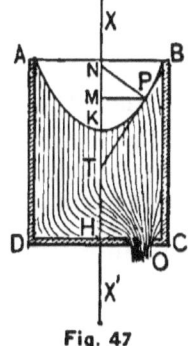

Fig. 47

Let h denote the height KH; then the velocity of efflux through an orifice at H $= \sqrt{2gh}$. Let y denote the distance HO = MP of an orifice O from the axis XX', and ω the angular velocity; then, since the subtangent MT is bisected at K, we have, for the height of the liquid at P above the centre K,

$$KM = \tfrac{1}{2}TM = \tfrac{1}{2}MP \tan MPT$$

$$= \tfrac{1}{2}y \frac{MP}{MN}$$

$$= \frac{y^2 \omega^2}{2g} \text{ [from (2) of Art. 21]}.$$

Hence, the velocity of efflux through the orifice at O is

$$v = \sqrt{2g\left(h + \frac{y^2\omega^2}{2g}\right)}$$
$$= \sqrt{2gh + y^2\omega^2}. \qquad (1)$$

Sch.—This formula is true for a vessel of any shape, even when it is closed at the top so that the paraboloid AKB cannot be completely formed. In this case also, h is the depth of the orifice below the vertex K, and $y\omega$ is the velocity of rotation of the orifice. (See Weisbach's Mechs., p. 819.)

90. The Clepsydra, or Water-Clock.—This is an instrument consisting merely of a vessel from which the water is allowed to escape through an orifice in the bottom, and the intervals of time are measured by the depressions of the upper surface. Thus, if we wish the clock to run 12 hours, we let $t = 12$ hours $= 12 \times 60 \times 60$ seconds; then solving (4) of Art. 79 for h, we have

$$h = \frac{1}{2}\frac{k^2 g}{K'^2} t^2; \qquad (1)$$

and substituting in it this value of t, we have

$$h = \frac{1}{2}\frac{k^2 g}{K'^2}(12 \times 60 \times 60)^2,$$

which gives the depth of liquid in the cylindrical vessel that will empty itself in 12 hours.

(1) To discover the manner in which the height h of the vessel must be divided in order that the upper surface of the liquid may descend through the several divisions of the scale in equal intervals of time, we make t in (1) successively equal to 12, 11, 10, ...4, 3, 2, 1 hours, and get for h a series of values which are as 144, 121, 100, ...16, 9, 4, 1; hence, if the height h be divided into 144 equal spaces, and

marked upwards from the bottom of the vessel, then the marks 121, 100, ... 16, 9, 4, 1, 0, will give the water level at 1, 2, ... 8, 9, 10, 11, 12 hours after the water begins to flow.

(2) Any vessel may serve for a clepsydra, but that form is most convenient in which the upper surface of the liquid descends uniformly.

Let $x =$ the height of the liquid in the vessel, K the area of the descending surface, v its velocity, and k the area of the orifice. Then from (a) of Art. 78, we have

$$v = \frac{k}{K}\sqrt{2gx}. \qquad (2)$$

And since the surface is to descend uniformly, this value of v must be equal to some constant a, which will depend upon the whole height and the time in which the clepsydra will be emptied; hence (2) becomes

$$K^2 = \frac{k^2\,2gx}{a^2}; \qquad (3)$$

and supposing the area of the descending surface of the liquid to be a circle $= \pi y^2$, (3) becomes

$$K^2 = \pi^2 y^4 = \frac{k^2\,2gx}{a^2};$$

$$\therefore\ y^4 = \frac{2k^2 g}{\pi^2 a^2}x, \qquad (4)$$

which is a parabola of the fourth order.

Hence, *the heights of the sections must vary as the fourth power of their radii.*

91. The Vena Contracta.—The laws of efflux that have been deduced are founded on the hypothesis that the liquid particles descend in straight lines to the orifice, and all issue in parallel lines with a velocity due to the height of the liquid surface. Experiment shows, however, that this is not the case. The liquid does not issue in the form of a prism, and hence the quantity discharged in a unit of time is not measured by the contents of a prism whose base is the orifice and whose altitude is the velocity; this would give the *theoretical* discharge (Art. 76, Cor. 3), but the *practical* discharge is generally much less. When a vessel empties itself through an orifice, it is observed that the particles of liquid near the top descend in vertical lines; but when they approach the bottom they take a curvilinear course, being turned in towards the orifice, or spirally around it, and this deviation from a vertical rectilinear path is the greater the further the horizontal distance of the particles is from the orifice. The oblique direction of the exterior particles within the vessel continues through the orifice, and gives the stream of liquid, in issuing from the orifice, nearly the form of a truncated cone or pyramid, whose larger base is the area of the orifice. This diminution in the size of the issuing stream is called the *contraction of the vein*, and the section of the stream at the point of greatest contraction is called the *Vena Contracta*,* or *contracted vein*.

From the results of most experiments, the *vena contracta*, when the orifice is a circle, is at a distance from the orifice equal to the radius of the orifice.

92. **Coefficient of Contraction.**—When water flows through orifices in thin plates, it has been found, by measurements of the stream, made by different experimenters, that its diameter at the *vena contracta* is about 0.8 of the diameter of the orifice. The ratio, therefore, of the cross-section of the *vena contracta* to that of the orifice in a thin

---

* This name was first given by Newton, who also showed that, by taking the area of the *rena contracta* as the area of the orifice, and regarding the height of the surface above the *vena contracta* as the height of the vessel, the theoretic discharge agreed far more closely with the practical.

plate is 0.64. This ratio is called the *Coefficient of Contraction*.* Denoting it by a, we have ak for the section of the *vena contracta*, k being the section of the orifice (Art. 76). Substituting ak for k in (6) of Art. 76, we have, for the quantity Q_1 discharged,

$$Q_1 = akv = ak\sqrt{2gh} = .64k\sqrt{2gh}, \qquad (1)$$

which is the quantity discharged in a unit of time.

93. Coefficient of Velocity.—The actual velocity of discharge is found by experiments to be a little less than the theoretical velocity, $v = \sqrt{2gh}$. Experiments † made with polished orifices have shown that the actual velocity is from 96 to 99 per cent. of the theoretical one. This loss of velocity arises from the friction of the water upon the inner surface of the orifice, and from the viscosity of the water. The ratio of the actual velocity to the theoretical velocity is called the *Coefficient of Velocity*. This coefficient is found to be tolerably constant for different heads with well-formed simple orifices, and it very often has the value 0.97. Denoting the coefficient of velocity by ϕ, and the actual velocity by v_1, we have

$$v_1 = \phi v = \phi\sqrt{2gh} = .97\sqrt{2gh}, \qquad (1)$$

which is the actual velocity of efflux.

94. Coefficient of Efflux.—If the value of v_1 in (1) of Art. 93 be substituted for the velocity $\sqrt{2gh}$ in (1) of Art. 92, we have, for the actual discharge Q_2,

$$\begin{aligned}Q_2 &= ak\phi v = ak\phi\sqrt{2gh} \\ &= .64 \times .97k\sqrt{2gh} = .62k\sqrt{2gh}.\end{aligned} \qquad (1)$$

---

* This ratio is not constant, but undergoes variations by varying the form of the orifice, the thickness of the surface in which the orifice is made, or the form of the vessel.

† Experiments made by Michelotti, Eytelwein, and others.

The ratio of the actual discharge Q_2 to the theoretical discharge Q is called the *Coefficient of Efflux*.

Denoting the coefficient of efflux by μ, we have, from (1) and (6) of Art. 76,

$$\mu = \frac{Q_2}{Q} = a\phi = .62 ; \qquad (2)$$

i. e., the coefficient of efflux is the product of the coefficient of velocity and the coefficient of contraction.

Sch.—The value of μ can also be determined by direct measurement of the discharge in a given time, an observation which can be made with much greater accuracy than those of contraction and velocity, on which it depends. In the present case it is found by direct measurement to be .62, agreeing well with the product $.64 \times .97$, of the values above given.*

Rem.—Repeated observations and experiments have led to the conclusion that the coefficient of efflux is not constant for all orifices in thin plates; that it is greater for small orifices and small velocities of efflux than for large orifices and great velocities, and that it is much greater for long narrow orifices than for those whose forms are regular or circular. For square orifices, whose areas are from 1 to 9 square inches, under a head of from 7 to 21 feet, according to the experiments of Bossut and Michelotti, the mean coefficient of efflux is $\mu = .610$; for circular orifices from $\frac{1}{2}$ to 6 inches in diameter, with from 4 to 20 feet head of water, it is $\mu = .615$, or about $\frac{8}{13}$.†

95. Efflux through Short Tubes, or Ajutages.—If the water, instead of flowing through an orifice in a thin plate, be allowed to discharge through *short tubes*, called also *ajutages* and *mouth-pieces*, the quantity discharged from a given orifice is considerably increased. More seems to be gained by the adhesion of the liquid particles to the sides of the tube, in preventing the contraction of the stream, than is lost by the friction. Ajutages of different forms have

---

* Cotterill's Applied Mechs., p. 449.
† Weisbach's Mechs., p. 824; also, Tate's Mech. Phil., p. 282.

different degrees of advantage in this respect, which can be determined only by experiment. The discharge is found to be greater when the ajutage is conical and the larger end is the discharging orifice.

(1) The results of many experiments * made with cylindrical tubes 1½ to 3 inches in diameter, the length of which does not exceed 4 times the diameter, as in Fig. 48, and under a head of water varying from 3 to 20 feet, give as a mean value of the coefficient of efflux, $\mu = .815$, or about ⅘. Since the coefficient of efflux for a simple orifice in a thin plate (Art. 94) is $\mu = .615$, it follows that, when the other circumstances are the same, the discharge through

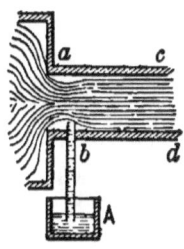

Fig. 48

a short cylindrical tube $= \dfrac{.815}{.615} = 1.325$ times the discharge through a simple orifice in a thin plate. These coefficients increase a little when the diameter of the tube becomes greater, and decrease a little when the head of water or the velocity of efflux increases.

In this tube, the contraction of the stream takes place at the inlet ab, and not at the outlet. If a small hole were bored in the tube at a or b, no water would run out, but air would be sucked in; and when the hole is enlarged, or when several of them are made, the discharge with a filled tube ceases. Also, if a tube be placed in a vessel of water A, and inserted in the hole at b, the water will rise in the tube Ab, and run out of the tube $abcd$.

(2) With a compound mouth-piece, having an enlargement at its exterior orifice or outlet, as well as at its interior orifice, as in Fig. 49, the results of careful experiments † give the coefficient of efflux $\mu = 1.5526$, when the narrow part cd is treated as the orifice, thus

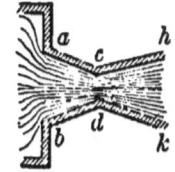

Fig. 49

---

* Experiments made by Michelotti.   † Made by Eytelwein.

giving a discharge greater than that which is due to the section cd of the pipe. Since $\mu = .615$ for a simple orifice, it follows that the discharge through the compound mouthpiece

$$= \frac{1.5526}{.615} = 2.5 \text{ times the discharge through a simple orifice in a thin plate,}$$

and $$= \frac{1.5526}{.815} = 1.9 \text{ times the discharge through a short cylindrical tube.}$$

In the experiments made by Eytelwein, the interior diameter ab was about 1.2 times the diameter cd, and the sides ch and dk made with each other an angle of 5° 9'.

96. Coefficient of Resistance. — When water flows from a cistern through a tube kept constantly full, it follows that the coefficient of contraction of this mouth-piece $a =$ unity, and that its coefficient of velocity $\phi =$ its coefficient of efflux μ.

Let W be the weight of water discharged with the actual velocity v, and v_1 the theoretical velocity of discharge due to the head of water h. Then the actual kinetic energy, or stored work, of the weight W of water, which issues with a velocity v,

$$= \frac{v^2}{2g} W \quad \text{(Anal. Mechs., Art. 217).} \quad (1)$$

But since the theoretical velocity of efflux $= v_1$, the theoretical kinetic energy or stored work of the weight W of water discharged

$$= \frac{v_1^2}{2g} W. \quad (2)$$

Hence, the loss of kinetic energy or stored work of the weight W of water discharged, during the efflux

$$= (v_1^2 - v^2) \frac{W}{2g}. \quad (3)$$

But, from (1) of Art. 93,

$$v = \phi v_1; \quad \therefore \ v_1 = \frac{v}{\phi},$$

which in (3) gives

$$\text{stored work lost} = \left(\frac{1}{\phi^2} - 1\right)\frac{v^2}{2g} W. \quad (4)$$

This loss of stored work corresponds to the head of water

$$\left(\frac{1}{\phi^2} - 1\right)\frac{v^2}{2g}, \quad (5)$$

which we can therefore consider as the loss of head due to the resistance of efflux, and we can assume that, when this loss has been subtracted, the remaining portion of the head is employed in producing the velocity.

The loss of head in (5), which varies as the square of the velocity, is known as the *height of resistance*.

The coefficient $\frac{1}{\phi^2} - 1$, by which the head of water $\frac{v^2}{2g}$ must be multiplied in order to obtain the height of resistance, *i. e.*, the ratio of the height of resistance to the head of water, is called the *Coefficient of Resistance*.

COR. 1.—Denote the coefficient of resistance by β; then we have

$$\beta = \frac{1}{\phi^2} - 1, \quad (6)$$

which in (5), and denoting the loss of head or the height of resistance by z, we have

$$z = \beta \frac{v^2}{2g}. \quad (7)$$

COR. 2.—For efflux through well-formed smooth orifices in a thin plate, the mean value of ϕ = the mean of .96 and .99 (Art. 93), = 0.975, and therefore we have, from (6),

$$\beta = \left[\left(\frac{1}{.975}\right)^2 - 1\right] = 0.052,$$

which in (4) gives for the loss of energy, or stored work lost,

$$0.052 \frac{v^2}{2g} W, \text{ or } 5.2 \text{ per cent.} \tag{8}$$

COR. 3.—For efflux through a short cylindrical tube [Art. 95, (1)], we have $\phi = .815$, since $\phi = \mu$, and therefore we have, from (6),

$$\beta = \left[\left(\frac{1}{.815}\right)^2 - 1\right] = 0.505,$$

which in (4) gives, for the loss of energy,

$$0.505 \frac{v^2}{2g} W, \tag{9}$$

or nearly 10 times as much as for efflux through an orifice in a thin plate.

SCH.—Hence, if the kinetic energy of the water is to be made use of, it is better to allow it to flow through an orifice in a thin plate than through a short cylindrical tube. But if the edge of the tube be rounded off where it is united to the interior surface of the vessel, and shaped like the contracted vein, we have $\mu = \phi = .975$, and the loss of energy is the same as it is for an orifice in a thin plate, *i. e.*, 5.2 per cent.

COR. 4.—From (6) we have

$$\phi = \frac{1}{\sqrt{1+\beta}}, \tag{10}$$

which gives the coefficient of velocity in terms of the coefficient of resistance.

EXAMPLE.

What is the discharge under a head of water of 3 feet through a tube 2 inches in diameter, whose coefficient of resistance is $\beta = 0.4$? Here from (10) we have

$$\phi = \frac{1}{\sqrt{1.4}} = 0.845;$$

hence, from (1) of Art. 93, we have, for the actual velocity v_1,

$$v_1 = \phi\sqrt{2gh}$$
$$= 0.845\sqrt{64.4 \times 3}$$
$$= 0.845 \times 8.025\sqrt{3} = 11.745 \text{ feet;}$$
$$k = (\tfrac{1}{12})^2 \pi = 0.02182 \text{ square feet;}$$

hence, the required discharge, from (1) of Art. 94 (since $\alpha = 1$) is

$$Q = k\phi\sqrt{2gh}$$
$$= 0.02182 \times 11.745 = 0.256 \text{ cubic feet.}$$

97. Resistance and Pressure of Fluids. — (1) By the *resistance of fluids* is meant that force by which the motions of bodies therein are impeded. The resistance of a fluid to the motion of a body is occasioned by the force necessary to displace that fluid. Since the motion communicated to a body at rest by another body impinging on it with a certain velocity is equal to the motion lost by the impinging body, therefore the motion communicated to the displaced fluid must be the same as that of the moving body; hence the work which the fluid destroys in the moving body will be equal to the work stored in the fluid.

Let $a =$ the area of the front of the body presented to the fluid, $v =$ the velocity of the body, $w =$ the weight of a cubic foot of the fluid, R = the resistance of the fluid to the motion of the body. Then,

weight of the displaced fluid per second = avw.

But this mass has a velocity of v feet given to it;

∴ work generated per second in displacing this fluid

$$= \frac{awv^3}{2g}. \qquad (1)$$

But this work is performed by means of a force which drags the body through the water at the rate of v feet per second, against an equal and opposite resistance R;

$$\therefore R \times v = \frac{awv^3}{2g};$$

$$\therefore R = \frac{awv^2}{2g}; \qquad (2)$$

that is, *the resistance varies as the square of the velocity*.

On account of eddies which are formed round the corners of the body and in the rear, the value of R in (2) should be multiplied by a constant k, giving

$$R = kaw\frac{v^2}{2g}. \qquad (3)$$

REM.—The constant k is to be determined by experiment for each form of solid. For a body whose transverse section is circular, k does not differ much from unity; for a flat plate moving flat-wise, it is about 1.25. Resistances of this kind, however, are very irregular, and may vary considerably in the course of the same experiment. Different results are therefore obtained by different experimentalists.*

---

* See Rankine's Applied Mechs., p. 598; also Cotterill's Applied Mechs., p. 472.

(2) The *pressure* of a current upon a plane is equal to the resistance suffered by the same plane when moving in the same direction and with the same velocity through the fluid; therefore (3) will also represent the pressure which the current, moving with the velocity v, would exert against the plane at rest. Calling F the pressure, we have

$$F = kaw\frac{v^2}{2g}. \qquad (4)$$

98. Work and Pressure of a Stream of Water.—

To find the work of a stream of water which impinges perpendicularly upon the surface of a heavy body which is itself in motion, and whose weight is very great as compared with that of the impinging water.

Let AB represent a plane surface moving horizontally with velocity v_1, while a horizontal jet moving with greater velocity v, strikes it centrally. Let W be the weight of water acting on the surface per second.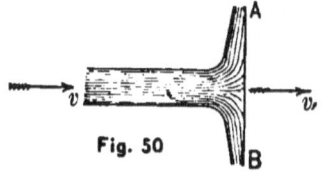

Fig. 50

Then the stored work or kinetic energy of the water

$$= \frac{v^2}{2g} W, \qquad (1)$$

and if the body were at rest, this would be the loss of energy.

From Anal. Mechs., Art. 208, (4), if m' be very great as compared with m, the loss of kinetic energy by impact becomes

$$\tfrac{1}{2} m (v - v_1)^2. \qquad (2)$$

Hence, if we first suppose that the water after impact moves on with the velocity of the body, we have by (2),

$$\text{work lost by impact} = (v - v_1)^2 \frac{W}{2g}. \qquad (3)$$

From (1) and (3) we have

$$\text{work done on the body} = \frac{v^2}{2g}W - (v-v_1)^2 \frac{W}{2g}$$

$$= [v^2 - (v-v_1)^2]\frac{W}{2g}. \quad (4)$$

Now if the water leaves the body, there will be more work lost, *i.e.*, the work remaining in the water will be lost; therefore we have

$$\text{work done on the body} = [v^2 - (v-v_1)^2]\frac{W}{2g} - \frac{v_1^2}{2g}W$$

$$= (v-v_1)v_1 \frac{W}{g}. \quad (5)$$

Cor. 1.—If P denote the pressure of the water against the body, then the work done on the body $= Pv_1$, which in (5) gives

$$P = (v-v_1)\frac{W}{g}. \quad (6)$$

If the body is at rest, or $v_1 = 0$, (6) becomes

$$P = \frac{W}{g}v. \quad (7)$$

Cor. 2.—Let $a =$ the section of the pipe, and $v =$ the velocity due to the head of water h; then $W = 62.5av$, which in (7) gives

$$P = 62.5a \times 2h. \quad (8)$$

EXAMPLE.

To find the work of a stream of water issuing from a nozzle with a given velocity.

Let v be the given velocity, a the area of the nozzle, and w the weight of a cubic foot of water. Then the weight of the water projected per second $= awv$, and therefore the work per second

$$= \frac{awv^3}{2g}; \qquad (1)$$

that is, *the work varies as the cube of the velocity of the water*.

Cor.—Let $\phi =$ the coefficient of velocity; then, from (1) of Art. 93, we have

$$v = \phi\sqrt{2gh},$$

which in (1) gives

$$\text{work per second} = \phi^3 awh\sqrt{2gh}. \qquad (2)$$

99. Impact of a Stream of Water against any Surface of Revolution.—Let BAC be a surface of revolution, against which a stream of water FA, moving in the direction of the axis AP of the surface, impinges. Let W be the weight of water discharged on the surface per second, v its velocity, v_1 the velocity of the surface, and α the angle BTP which the tangent HT to the surface at B makes with the axis

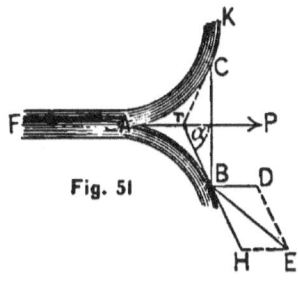

Fig. 51

AP, or which each filament HB of the stream of water, on leaving the surface, makes with the direction of the axis BD. Then the water impinges upon the surface with the velocity $v - v_1$; and, if friction be neglected, the water passes over the surface with that velocity, and leaves it in a tangential direction, TH, TK, etc., with the same velocity. From the tangential velocity $BH = v - v_1$, and the velocity BD $= v_1$ of the surface parallel to the axis, we have the resultant velocity $BE = V$ of the water, after it has impinged on the surface, by the formula for the parallelogram of velocities,

$$V = \sqrt{(v - v_1)^2 + v_1^2 + 2(v - v_1)v_1 \cos \alpha}. \quad (1)$$

Now the kinetic energy of the water before impact

$$= \frac{v^2}{2g} W,$$

and the kinetic energy remaining in the water after impact

$$= \frac{V^2}{2g} W; \quad (2)$$

hence, the kinetic energy transmitted to the surface

$$= (v^2 - V^2) \frac{W}{2g}. \quad (3)$$

If P be put for the force or impulse against the surface, then the energy transmitted to the surface $= Pv_1$, which in (3) gives

$$Pv_1 = (v^2 - V^2) \frac{W}{2g}$$

$$= [v^2 - (v - v_1)^2 - v_1^2 - 2(v - v_1)v_1 \cos \alpha] \frac{W}{2g},$$

from (1).

$$= (1 - \cos \alpha)(v - v_1) v_1 \frac{W}{g}; \quad (4)$$

$$\therefore P = (1 - \cos \alpha)(v - v_1) \frac{W}{g}, \quad (5)$$

which is the force of the water against the surface in the direction of the axis.

That is, *the impulse varies as the relative velocity of the water.*

Cor. 1.—If the surface moves with a velocity v_1 in the *opposite direction* to that of the water, we have, from (5),

$$P = (1 - \cos \alpha)(v + v_1) \frac{W}{g}. \quad (6)$$

If the surface is at rest, $v_1 = 0$, and (6) becomes

$$P = (1 - \cos \alpha) v \frac{W}{g}. \qquad (7)$$

Cor. 2.—If $a =$ the area of the cross-section of the stream, and $w =$ the weight of a cubic foot of the water, the weight of the impinging water per second is

$$W = (v \mp v_1) aw, \qquad (8)$$

which in (5) and (6) gives

$$P = (1 - \cos \alpha)(v \mp v_1)^2 \frac{aw}{g}, \qquad (9)$$

and in (7) gives $P = (1 - \cos \alpha) v^2 \dfrac{aw}{g}.$ \qquad (10)

That is, *the impulse varies as the square of the relative velocity of the water, and also as the area of the cross-section of the stream.*

Cor. 3.—The impulse of the same stream of water depends principally upon the angle α at which the water moves off from the axis after the impact. If the surface BAC is hollow, as in Fig. 52, the water after impact leaves the surface in a direction opposite to that in which it strikes it, and thus much more work is done on the body with a surface concave to the stream than on one convex to the stream, since the work remaining in the water on leaving the former surface will be less than it is in the water on leaving the latter. If $\alpha = 180°$, we have $\cos \alpha = -1$, which in (5) and (6) gives

Fig. 52

$$P = 2(v \mp v_1)\frac{W}{g}, \qquad (11)$$

and in (7) gives
$$P = 2v\frac{W}{g}. \quad (12)$$

COR. 4.—When the surface is plane, as in Fig. 50, $\alpha = 90°$ and $\cos \alpha = 0$. Substituting this value in (5), (6), and (7), they become

$$P = (v \mp v_1)\frac{W}{g}, \quad (13)$$

which agrees with (6) of Art. 98; and

$$P = v\frac{W}{g} = \frac{v^2}{g}aw, \text{ from (8)},$$

$$= 2 \times \frac{v^2}{2g} \times aw = 2h \times aw; \quad (14)$$

that is, *the normal impulse of water against a plane surface is equal to the weight of a column of water whose base is equal to the cross-section of the stream, and whose height is twice the head of water to which the velocity is due.*

COR. 5.—If the plane surface (Fig. 50) against which the stream impinges moves away with a velocity u in a direction which makes an angle θ with the original direction of the stream, the velocity of the surface in the direction of the impact is

$$v_1 = u \cos \theta,$$

which in (13) gives for the impulse,

$$P = (v - u \cos \theta)\frac{W}{g}, \quad (15)$$

and the work done by it per second is

$$Pv_1 = (v - u \cos \theta) u \cos \theta \frac{W}{g}. \quad (16)$$

100. Oblique Impact.

When a stream impinges obliquely on a plane, there are several cases, viz., when the water after impact flows off in one, two, or in more directions.

(1) Let the plane AB, upon which the stream AC impinges, have a border upon three sides so that the water can flow off in one direction only.

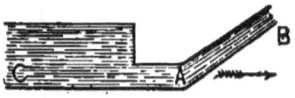

Fig. 53

Then the impulse of the water against the surface in the direction of the stream is, from (5) of Art. 99,

$$P = (1 - \cos \alpha)(v - v_1)\frac{W}{g}. \qquad (1)$$

(2) Let the plane AB, upon which the stream DC impinges, have a border upon two sides only, so that the water can flow off in only two directions. The stream will divide itself into two unequal parts, the greater part flowing off

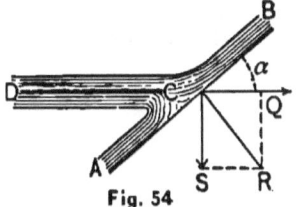

Fig. 54

in the direction CB, and the other in the direction CA. Let W_1 be the weight of the former, W_2 the weight of the latter, and W the whole weight. Then the total impulse in the direction of the stream, from (5) of Art. 99, is

$$P = (1 - \cos \alpha)(v - v_1)\frac{W_1}{g} + (1 + \cos \alpha)(v - v_1)\frac{W_2}{g}$$

$$= \left(\frac{v - v_1}{g}\right)[(1 - \cos \alpha) W_1 + (1 + \cos \alpha) W_2]. \qquad (2)$$

But the conditions of equilibrium of the two portions of the stream require that the pressures on CB and CA shall be equal to each other; hence

$$(1 - \cos \alpha) W_1 = (1 + \cos \alpha) W_2,$$

or, $\qquad (1 - \cos \alpha) W_1 = (1 + \cos \alpha)(W - W_1),$

from which we find $W_1 = \frac{1}{2}(1 + \cos \alpha) W$,

and $\qquad W_2 = \frac{1}{2}(1 - \cos \alpha) W$.

Substituting these values of W_1 and W_2 in (2), we have

$$P = \frac{v - v_1}{g} W \sin^2 \alpha, \qquad (3)$$

which is the total impulse in the direction of the stream.

Dividing (3) by $\sin \alpha$, we obtain

$$\text{CR} = P \csc \alpha = \frac{v - v_1}{g} W \sin \alpha, \qquad (4)$$

which is the normal impulse.

Multiplying (3) by $\cot \alpha$, we obtain

$$\text{CS} = P \cot \alpha = \frac{v - v_1}{g} W \sin \alpha \cos \alpha$$

$$= \frac{v - v_1}{2g} W \sin 2\alpha, \qquad (5)$$

which is the lateral impulse.

Hence, *the total impulse in the direction of the stream is proportional to the square if the sine of the angle of incidence, the normal impulse to the sine of this angle, and the lateral impulse to the sine of double this angle.*

Sch.—If the oblique plane has no border, the water can flow off in all directions; in this case the impulse is increased, for α is the smallest angle which the filaments of water can make with the axis, and hence every filament which does not flow off in the normal plane will make with the axis an angle larger than α, and therefore from (3) will exert a greater pressure than those which do.

101. Maximum Work done by the Impulse.—

The work done by the impulse P, from (4) of Art. 99, is

$$Pv_1 = (1 - \cos \alpha)(v - v_1) v_1 \frac{W}{g}. \qquad (1)$$

The work is zero when the velocity of the surface $v_1 = 0$, and also when it $= v$.

To find the value of v_1 which makes this work a maximum, we must equate to zero its derivative with respect to v_1, which gives

$$v - 2v_1 = 0, \quad \text{or} \quad v_1 = \tfrac{1}{2}v;$$

hence, *the work done by the impulse is a maximum when the surface moves in the direction of the stream, with half the velocity of the stream.*

Substituting in (1) for v_1 its value, we have

$$Pv_1 = (1 - \cos \alpha) \frac{1}{2} \frac{v^2}{2g} W, \qquad (2)$$

which is the maximum work done by the impulse.

Cor.—If the surface is a plane, as in Fig. 50, $\alpha = 90°$, and we have, from (2),

$$Pv_1 = \frac{1}{2} \frac{v^2}{2g} W. \qquad (3)$$

That is, *the water transmits to the surface, in this case, one-half of its kinetic energy.*

If the surface is hollow, as in Fig. 52, so that the water is reversed, $\alpha = 180°$, and we have, from (2),

$$Pv_1 = \frac{v^2}{2g} W. \qquad (4)$$

In this case, the water transmits to the surface all of its kinetic energy. (See Weisbach's Mechs., p. 1010.)

EXAMPLES.

1. With what velocity will water issue from a small orifice $64\frac{1}{3}$ feet below the surface of the liquid?

Ans. $64\frac{1}{3}$ feet.

2. If 252 cubic inches of water flow in one second through an opening of 6 square inches, find the head of water.

Ans. 2.28 inches.

3. If water flows from a vessel whose cross-section is 60 square inches, through a circular orifice in the bottom 5 inches in diameter under a head of water of 24 feet, find its velocity. *Ans.* 41.58.

4. A vessel, formed by the revolution of a semi-cubical parabola about its axis, which is vertical, is filled with water till the radius of its surface is equal to its height above the vertex. Find the time of emptying the vessel through a small orifice at the vertex.

[Let $ay^2 = x^3$ be the equation of the generating curve, and k the area of the orifice.]

Ans. $\dfrac{\pi a^2}{7k}\sqrt{\dfrac{2a}{g}}$.

5. A conical vessel, the radius of whose base is r and altitude h, is filled with water; the axis is vertical and the water issues through an orifice in the vertex, of area k. Find (1) the time in which the surface of the water will descend through one-half its altitude, and (2) the time in which the cone will empty itself.

Ans. (1) $\dfrac{\pi r^2}{20k}(2^{\frac{5}{2}} - 1)\sqrt{\dfrac{h}{g}}$; (2) $\dfrac{2\pi r^2}{5k}\sqrt{\dfrac{h}{2g}}$.

6. Find the time in which the cone in Ex. 5 would empty itself through an orifice in its base.

Ans. $\dfrac{16\pi r^2}{15k}\sqrt{\dfrac{h}{2g}}$.

7. A sphere is filled with water. Find the time of emptying it through an orifice in its bottom. *Ans.* $\dfrac{16\pi r^2}{15k}\sqrt{\dfrac{r}{g}}$.

8. A hemisphere is filled with water. Find the time of emptying it (1) through an orifice in its vertex, and (2) through an orifice in its base.

$$Ans. \quad (1) \ \frac{14\pi r^{\frac{5}{2}}}{15k\sqrt{2g}}; \quad (2) \ \frac{8\pi r^{\frac{5}{2}}}{5k\sqrt{2g}}.$$

9. A rectangular orifice is 3 feet wide and $1\frac{1}{4}$ feet high, and the lower edge is $2\frac{3}{4}$ feet below the level of the water. Find the quantity discharged in 1 second.

Ans. 43.7 cubic feet.

10. An orifice in the form of an isosceles triangle, with its vertex in the surface of the water has a base of 1 foot which is horizontal, and an altitude of 6 inches. Find the quantity discharged in 1 second. *Ans.* 1.135 cubic feet.

11. If the orifice in Ex. 10 has its vertex downwards and its base 6 inches below the surface of the water, and horizontal, find the quantity discharged in 1 second.

Ans. 1.632 cubic feet.

12. If a vessel, when filled with water to the depth of 4 feet, weighs 350 lbs., and if it be drawn upwards by a weight P of 450 lbs., as in Fig. 46, find the velocity of efflux through an orifice in the bottom. *Ans.* 17.02 feet.

13. If the vessel (Fig. 47), which is filled with water, makes 100 revolutions per minute, and if the orifice O is 2 feet below the surface of the water at the centre, and at a distance of 3 feet from the axis XX', find the velocity of efflux. *Ans.* 33.4 feet.

14. Find the times in which the surface of water contained in a vessel, formed by the revolution of the curve $y^4 = a^3 x$ about the axis of x, will descend through equal distances h, the water issuing through a small orifice in the vertex, and the axis vertical.

$$Ans. \ \frac{\pi a^{\frac{3}{2}} h}{k\sqrt{2g}}.$$

15. Water issues through a small orifice $16\frac{1}{12}$ feet below the surface of the liquid. If the area of the orifice is 0.1 of

a square foot and the coefficient of efflux is 0.615, how many cubic feet of water will be discharged per minute?
Ans. 118.695.

16. A basin has in it a hole an inch square; water in the basin is kept at a constant level of 9 feet above the hole. How many cubic feet of water will flow out in 1 hour, the coefficient of efflux being 0.6 ? *Ans.* 360.

17. A cylindrical vessel filled with water is 4 feet high and 1 square foot in cross-section, and a hole of 1 square inch is made in the bottom. If the coefficient of efflux is 0.6, in what time will $\frac{3}{4}$ of the water be discharged?
Ans. 60 seconds, nearly.

18. A cylinder, the area of whose cross-section is 60 sq. ft., is filled with water to a depth of 12 feet; a small hole is made in its bottom, whose area is 0.5 square inches. In how long a time will the depth of the water be (1) 8 feet and (2) 4 feet? *Ans.* (1) 45.8 minutes; (2) 105.4 minutes.

19. The horizontal section of a cylindrical vessel is 100 square inches, its altitude is 36 inches, and the area of its orifice is 0.1 of a square inch. If filled with water, in what time will it empty itself, the coefficient of efflux being 0.62 ? *Ans.* 11 m. 36.5 s.

20. What is the discharge per second through a rectangular orifice 2 feet wide and 1 foot high, when the surface of the water is 15 feet above the upper edge, the coefficient of efflux being 0.611 ? *Ans.* 38.6 cubic feet.

21. What is the discharge per second through a rectangular orifice whose height is 8 inches and whose width is 2 inches, under a head of water of 15 inches above the upper edge, the coefficient of efflux being 0.628 ?
Ans. 0.705 cubic feet.

22. If the height of the rectangular orifice is 15 inches, its width 25 inches, and the head of water is 4½ inches above the upper edge, what is the discharge per second, the coefficient of efflux being 0.594 ? *Ans.* 12.19 cubic feet.

23. A plane area moves perpendicularly through water in which it is deeply imbedded. Find the resistance per square foot at a speed of 10 miles an hour. *Ans.* 269 lbs.

24. A stream of water delivering 100 cubic feet per minute, at a velocity of 15 feet per second, strikes an indefinite plane normally. Find the pressure on the plane.
Ans. 48.6 lbs.

25. If a stream of water, the area of whose cross-section is 64 square inches, impinges with a velocity of 40 feet per second against the convex surface of an immovable cone, in the direction of its axis, the vertical angle of the cone being 100°, find the impulse. *Ans.* 492.16 lbs.

26. A stream of water, the area of whose cross-section is 40 square inches, delivers 5 cubic feet per second, and strikes normally against a plane surface, which moves away with a velocity of 12 feet per second. Find (1) the impulse, (2) the maximum work, and (3) the maximum impulse.
Ans. (1) 58.125 lbs.; (2) 784.688 ft.-lbs.; (3) 87.19 lbs.

CHAPTER II.

MOTION OF WATER IN PIPES AND OPEN CHANNELS.

102. Resistance of Friction. — When a thin plate with sharp edges, completely immersed in water, is moving edgeways through the water, a certain resistance is experienced, which must be overcome by an external force. This resistance acts along tangentially between the plate and the water, and so far is analogous to the friction between solid surfaces, but it follows quite different laws, which have been obtained from many observations and experiments, and which may be stated as follows:*

(1) The resistance of friction is entirely independent of the pressure on the surface.

(2) It varies as the area of the surface in contact with the water.

(3) It varies nearly as the square of the velocity. †

Hence, if R be the resistance of friction, S the area of the surface, and v the velocity, these laws may be expressed by the formula,

$$R = f S v^2, \qquad (1)$$

where f is called the "coefficient of friction," as in the friction of solid surfaces. The value of f depends on the smoothness of the surface; thus, for thin boards, with a clean, varnished surface, moving through water, it is found

---

* Cotterill's App. Mechs., p. 458.

† At low velocities, of not more than 1 inch per second for water, the resistance varies nearly as the first power of the velocity. At velocities of ¼ foot per second, and greater velocities, the resistance varies more nearly as the square of the velocity.

to be .004, while for a surface resembling medium sandpaper, it is .009, the units being pounds, feet, and seconds.*

103. Motion of Water in Pipes. — When water is conveyed to any considerable distance in pipes, the friction of the internal surface causes a great resistance to the flow. By the theoretical rule, the velocity of discharge v would be due to the vertical depth h through which the water falls (Art. 76); but owing to friction, theoretical results are of very little practical value. Besides, the friction is often quite uncertain, the central parts of the stream move more quickly than the parts in immediate contact with the pipe, and, though the circumstances are different, the velocity over the internal surface is liable to changes, as in the case of solid surfaces. The value of f therefore has to be obtained by special experiments, and the results of such experiments do not always agree with each other. It is found, however, that f lies between the limits .005 and .01, depending partly on the condition of the internal surface, and partly on the diameter and velocity; its value being greater in small pipes than in large ones, and greater at low velocities than at high ones. The mean of these limits, or .0075, is sometimes taken for f, when there is no special cause for increased resistance.

Let $v =$ the velocity of discharge in feet per second, $d =$ the diameter of the pipe in feet, $l =$ the length of the pipe in feet, $h =$ the head or fall of water in feet, and $W =$ the weight of water in pounds discharged per second. Let f' be the resistance of friction due to a unit of diameter, length, and velocity; then the resistance in a pipe l feet long and d feet diameter with a unit of velocity will be, from (1) of Art. 102, $f'ld$; but the quantity of water deliv-

---

* For large surfaces, especially of considerable length, the friction is very much diminished. For instance, these values of f were obtained by experimenting on a surface 4 feet long, moving 10 feet per second: but when the length was 20 feet and upwards, these values of f were diminished to .0025 and .005 respectively.

ered by this pipe will be d^2 times that delivered by the former, therefore for the same quantity of water delivered as by the former, the resistance of friction in the latter pipe will be

$$\frac{f'ld}{d^2} \quad \text{or} \quad f'\frac{l}{d};$$

that is, *the resistance of friction in pipes, when the velocity is constant, varies directly as their lengths and inversely as their diameters.*

If we measure this resistance by a column of water, and denote the height of this column by h_1, we have

$$h_1 = f\frac{l}{d}\frac{v^2}{2g}, \qquad (1)$$

where f is a constant to be determined by experiment, and is called the *coefficient of friction*.

This height h_1 is called *the height of resistance of friction*, which has to be subtracted from the total head h, in order to obtain the height necessary to produce the velocity v. Hence, the loss of head or of pressure, in consequence of the friction of the water in the pipe, is found by multiplying the head due to the velocity by the coefficient $f\frac{l}{d}$, and is greater, the greater the ratio of the length to the diameter and the greater the height due to the velocity.

Multiplying (1) by W, we obtain for the work due to the resistance of friction

$$h_1 W = f\frac{l}{d}\frac{v^2}{2g} W; \qquad (2)$$

that is, *the loss of work by friction is the same as that of raising the water through a height h_1.*

Cor.—From (4) of Art. 96, we have

loss of work due to the resistance at ingress $= \beta \dfrac{v^2}{2g} W$; (3)

work stored in the water at discharge $= \dfrac{v^2}{2g} W$. (4)

104. Uniform Pipe connecting Two Reservoirs, when all the Resistances are Considered.—Let h be the difference of level of the reservoirs, and v the velocity, in a pipe of length l and diameter d. Then we have

work due to the head of water $= hW$, (1)

which is the whole work done per second in moving W pounds of water from the surface of one reservoir to the surface of the other. This work is equal to the work in overcoming all the resistances, together with the work remaining in the water at discharge. That is, the work is expended in three ways: (1) The head $\dfrac{v^2}{2g}$,* corresponding to an expenditure of $\dfrac{v^2}{2g} W$ foot-pounds of work, is employed in giving energy of motion to the water, and is ultimately wasted in eddying motions in the lower reservoir. (2) A portion of head $\beta \dfrac{v^2}{2g}$, corresponding to an expenditure of $\beta \dfrac{v^2}{2g} W$ foot-pounds of work, is employed in overcoming the resistance at the entrance to the pipe. (3) the head †$f \dfrac{l}{d} \dfrac{v^2}{2g}$, corresponding to an expenditure of $f \dfrac{l}{d} \dfrac{v^2}{2g} W$ foot-pounds of work, is employed in overcoming the surface friction of the pipe. Hence, from (1), and (2), (3), (4) of Art. 103, we have

$$hW = f \dfrac{l}{d} \dfrac{v^2}{2g} W + \beta \dfrac{v^2}{2g} W + \dfrac{v^2}{2g} W;$$

---

* Called *velocity head*.   † Called *friction head*.

$$\therefore\ h = \left(1 + \beta + f\frac{l}{d}\right)\frac{v^2}{2g}, \qquad (2)$$

and
$$v = \sqrt{\frac{2gh}{1 + \beta + f\dfrac{l}{d}}}, \qquad (3)$$

or
$$v = 8.025\sqrt{\frac{hd}{(1+\beta)\,d + fl}}, \qquad (4)$$

where the constants β and f are to be determined by experiment.

When v and d are given, (2) is used to determine h; when h and d are given, (4) is used to determine v.

COR. 1.—If the pipe is bell-mouthed, β is about .08. If the entrance to the pipe is cylindrical, $\beta = 0.505$. Hence, $1 + \beta = 1.08$ to 1.505. In general, this is so small compared with $f\dfrac{l}{d}$ that for practical calculations it may be neglected; i. e., the losses of head, except the loss in surface friction, are neglected. It is only in short pipes and at high velocities that it is necessary to take account of the term $(1+\beta)$. For instance, in pipes for the supply of turbines, v is usually limited to 2 feet per second, and the pipe is bell-mouthed. In this case, we have

$$(1 + \beta)\frac{v^2}{2g} = 1.08 \times 4 \times .0155 = 0.067 \text{ foot.}^*$$

In pipes for the supply of towns, v may range from 2 to $4\frac{1}{2}$ feet per second, and then we have

$$(1 + \beta)\frac{v^2}{2g} = 0.1 \text{ to } 0.5 \text{ foot.}$$

In either case, this amount of head is small compared

---

* $\dfrac{1}{2g} = .0155$ foot.

with the whole fall in the cases which most commonly occur.

Cor. 2.—For very long pipes, $1+\beta$ is so small compared with $f\dfrac{l}{d}$, that it may be neglected altogether, and (2), (3), and (4) become

$$h = f\frac{l}{d}\frac{v^2}{2g}, \tag{5}$$

$$v = \sqrt{\frac{2ghd}{fl}}, \tag{6}$$

$$v = 8.025\sqrt{\frac{hd}{fl}}. \tag{7}$$

Using the value of f, as determined by Poncelet, viz., $f = .023$, with the value of $\beta = .5$, we have, from (3),

$$v = 47.9\sqrt{\frac{hd}{l + 54d}}. \tag{8}$$

Eytelwein gave a formula which nearly coincides with this. (See Storrow on Water Works, p. 56.)

When the pipe is very long, d is very small compared with l, and (8) becomes

$$v = 47.9\sqrt{\frac{hd}{l}}. \tag{9}$$

Rem.—It is immaterial as regards the velocity, and the quantity discharged, whether the pipe is horizontal or inclined upwards or downwards, so long as the length of the pipe and the total head, or depth of the end of discharge below the level of the surface of the water in the reservoir, remain unchanged. If the inclined pipe is longer than the horizontal, of course its surface will present more friction against the motion of the water than the horizontal one, and thus diminish the velocity of discharge; but if the inclined pipe be the same length as the horizontal, and have the same head, then each of them will discharge the same quantity in the same time.

It is evidently necessary, in every case, that the entrance to the pipe from the reservoir be placed sufficiently far below the water surface of the reservoir to allow the water to flow from the reservoir *into* the pipe, as fast as it afterwards flows *along* or *through* the length of the pipe to the end of discharge. For there must be at least sufficient head to overcome the resistance at the entrance to the pipe, and to allow the water in the reservoir to flow out of an opening freely into the air with that velocity which previous calculation shows it will have in the pipe. The remainder of the head, which is employed in overcoming the resistance of friction, and perhaps other resistances which will be considered hereafter, may be obtained by having the pipe incline downwards.

Since the friction in pipes of the same diameter increases as their lengths, when the water first enters the pipe it is opposed by but little friction, and has great velocity; but this velocity gradually diminishes as the advancing water meets the friction along increased lengths of the pipe, and finally becomes least when the water fills the whole length and begins to flow from the end of discharge. The velocity then becomes uniform along the pipe, and will continue to be so, if the velocity head and head due to the resistance at the entrance to the pipe are together sufficient to allow the water of the reservoir to enter the pipe with this same velocity.

105. Coefficient of Friction for Pipes Discharging Water.

—From the average of a great many experiments, the value of f for ordinary pipes is

$$f = 0.030268. \qquad (1)$$

But practical experience shows that no single value can be taken applicable to very different cases. The coefficient of friction, like the coefficient of efflux, is not perfectly constant. It is greater for low velocities than for high ones, *i. e.*, the resistance of friction of the water in pipes does not increase exactly as the square of the velocity. Prony and Eytelwein assumed that the loss of head by the resistance of friction increases with the first power of the velocity and with its square; and hence they established for this loss of head the formula

$$h_1 = (a_1 v + a_2 v^2) \frac{l}{d}, \qquad (2)$$

in which a_1 and a_2 denote constants to be determined by experiment.

In order to determine these constants, these authors availed themselves of 51 experiments made at different times by Couplet, Bossut, and du Buat upon the flow of water through long pipes. From these 51 experiments, the following numerical values were obtained:

Prony obtained, $a_1 = 0.0000693,$ $a_2 = 0.0013932.$
Eytelwein, $a_1 = 0.0000894,$ $a_2 = 0.0011213.$
D'Aubuisson,* $a_1 = 0.0000753,$ $a_2 = 0.0013700.$

Taking the value of h_1, and substituting it in (2) of Art. 104, instead of the value of h_1 as given in (1) of Art. 103, we have

$$h = (1 + \beta)\frac{v^2}{2g} + (a_1 v + a_2 v^2)\frac{l}{d}. \qquad (3)$$

Putting $\dfrac{1+\beta}{2g} = b$, and reducing, (3) becomes

$$hd = bdv^2 + a_2 lv^2 + a_1 lv, \qquad (4)$$

from which the value of v may be found. But the following method of approximating to the value of v is more convenient. From (4) we have

$$v\left(1 + \frac{a_1 l}{bd + a_2 l} \cdot \frac{1}{v}\right)^{\frac{1}{2}} = \sqrt{\frac{hd}{bd + a_2 l}};$$

Expanding the first member by the binomial theorem, and neglecting all the terms of the expansion after the second, since a_2 is considerably greater than a_1, we have

$$v\left[1 + \frac{a_1 l}{2(bd + a_2 l)} \cdot \frac{1}{v}\right] = \sqrt{\frac{hd}{bd + a_2 l}};$$

---

* Weisbach's Mechs., p. 866.

$$\therefore v = \sqrt{\frac{hd}{bd + \alpha_2 l} - \frac{\alpha_1 l}{2(bd + \alpha_2 l)}}. \quad (5)$$

Now if the pipe is cylindrical, $\beta = 0.505$, from Cor. 3 of Art. 69, and therefore we have

$$b = \frac{1 + \beta}{2g} = \frac{1.505}{64\tfrac{1}{3}}$$
$$= .0234,$$

and taking $\alpha_1 = .00007$ and $\alpha_2 = .00042$,* and substituting these values in (5) and reducing, we have

$$v = \sqrt{\frac{2380hd}{l + 54d} - \frac{l}{12(l + 54d)}}. \quad (6)$$

Cor.—When h is not very small, the last term of (6) may be neglected, and we have

$$v = \sqrt{\frac{2380hd}{l + 54d}}, \quad (7)$$

which is very nearly the same as (8) of Art. 104.

When the pipe is very long, d is very small compared with l, and (6) becomes

$$v = \sqrt{\frac{2380hd}{l} - \frac{1}{12}}. \quad (8)$$

When d is expressed in inches and all the other dimensions in feet, (8) of Art. 104 becomes

$$v = \sqrt{\frac{191.2hd}{l + 4.5d}}. \quad (9)$$

Sch.—The following short table gives Weisbach's values of the coefficient of friction for different velocities in feet per second: †

---

* Tate's Mech. Phil., p. 293.
† Ency. Brit., Art. *Hydromechanics*.

194   THE QUANTITY DISCHARGED FROM PIPES.

$v =$	0.1	0.2	0.3	0.4	0.5	0.6	0.7	0.8	0.9
$f =$	.0686	.0527	.0457	.0415	.0387	.0365	.0349	.0336	.0325
$v =$	1	1¼	1½	2	3	4	6	8	12
$f =$	.0315	.0297	.0284	.0265	.0243	.0230	.0214	.0205	.0193

EXAMPLE.

The length of a water-pipe is 5780 feet, the head of water is 170 feet, and the diameter of the pipe is 6 inches. Required the velocity of discharge.

By (6), we have

$$v = \sqrt{\frac{2380 \times 170 \times .5}{5780 + 54 \times .5}} - \frac{5780}{12(5780 + 54 \times .5)} = 5.81.$$

By (8), we have

$$v = \sqrt{\frac{2380 \times 170 \times .5}{5780}} - \frac{1}{12} = 5.82.$$

By (8) of Art. 104, we have

$$v = 47.9\sqrt{\frac{170 \times .5}{5780 + 54 \times .5}} = 5.8 \text{ feet, nearly.}$$

By (9) of Art. 104, we have

$$v = 47.9\sqrt{\frac{170 \times .5}{5780}} = 5.8.$$

It will be observed that these results are very nearly the same.

106. The Quantity Discharged from Pipes.—Let Q be the number of cubic feet discharged per second; then Q is given by the formula

$$Q = \frac{\pi}{4} d^2 v = 0.7854 d^2 v ; \tag{1}$$

and on substituting the value of v obtained from (1) of Art. 103, this becomes

$$Q = \frac{\pi}{4}\sqrt{\frac{2gh_1}{lf}} d^{\frac{5}{2}}, \tag{2}$$

which gives the value of Q in cubic feet per second, since all the dimensions are in feet.

If we require the number of gallons discharged per minute for a diameter of d inches, (1) becomes

$$G = C\sqrt{\frac{h_1}{l}} d^{\frac{5}{2}}, \tag{3}$$

where C is a constant whose value for $f = .03$ is 30, but which is often taken somewhat less (say 27), to allow for contingencies.*

Assuming that 1 cubic foot $= 6.2322$ gallons, we have, from (1), for the number of gallons discharged in 24 hours,

$$Q = \frac{\pi}{576} d^2 v \times 86400 \times 6.2322$$

$$= 2936.86 d^2 v. \tag{4}$$

From (1), we have

$$v = \frac{4Q}{\pi d^2} = 1.2732 \frac{Q}{d^2}, \tag{5}$$

which, in (1) of Art. 103, gives

$$h_1 = \frac{f}{2g} \left(\frac{4}{\pi}\right)^2 \frac{lQ^2}{d^5} ; \tag{6}$$

that is, *the height of resistance of friction in pipes varies inversely as the fifth power of the diameters, and directly as the length of the pipes.*

---

* See Cotterill's App. Mechs., p. 462.

Hence, if we wish to conduct a given quantity of water through a pipe with as little loss of head as possible, we must make the pipe as short and its diameter as large as we can. If the diameter of one pipe is double that of another, the friction in the former is $\frac{1}{32}$ of that in the latter.

Cor.—Putting $1 + \beta = 1.505$, and $\dfrac{1}{2g} = 0.0155$, we have, from (2) and (4) of Art. 104,

$$h = \left(1.505 + f \cdot \frac{l}{d}\right) 0.0155 v^2, \qquad (7)$$

and
$$v = 8.025 \sqrt{\frac{h}{1.505 + f \cdot \dfrac{l}{d}}}. \qquad (8)$$

EXAMPLES.

1. How many gallons of water would the pipe in the example of Art. 105 deliver in 24 hours?

Here $v = 5.8$ and $d = 6$ inches; we have, from (4),

$$Q = 2936.86 \times 6^2 \times 5.8$$
$$= 613216 \text{ gallons in 24 hours.}$$

2. What must the head of water be, when a set of pipes 150 feet long and 5 inches in diameter is required to deliver 25 cubic feet of water per minute?

Here we have, from (5),

$$v = 1.2732 \times \frac{25}{60} \times \frac{12^2}{5^2} = 3.056 \text{ feet,}$$

and therefore (Art. 105, Sch.), $f = .0243$, which in (7) gives

$$h = \left(1.505 + .0243 \times \frac{150 \times 12}{5}\right).0155 \times \overline{3.056}^2$$

$$= (1.505 + 8.748).0155 \times 9.339 = 1.484 \text{ feet.}$$

3. Solve Ex. 1 by using the value of v as obtained from (8).

From (8), we have

$$v = 8.025\sqrt{\dfrac{170}{1.505 + f\dfrac{5780}{.5}}}.$$

Since v is somewhere between 3 and 10, we assume $f = .02$, and obtain

$$v = 8.025\sqrt{\dfrac{170}{1.505 + 231.20}}$$
$$= 6.859.$$

But $v = 6.9$ gives more correctly (Art. 105, Sch.) $f = .021$, and therefore we have

$$v = 8.025\sqrt{\dfrac{170}{1.505 + 244.265}}$$
$$= 6.695,$$

which gives the true value to the first decimal place.

The discharge, from (4), is

$$Q = 2936.86 \times 6^2 \times 6.7$$
$$= 708370.632 \text{ gallons in 24 hours.}$$

This result is somewhat larger than that obtained from the value of v in (8) of Art. 104.

107. The Diameter of Pipes.—Substituting in (1) of Art. 106 the value of v given in (9) of Art. 105, we have

$$Q = \dfrac{\pi}{576} d^2 \sqrt{\dfrac{191.2hd}{l + 4.5d}};$$

EXAMPLE.

$$\therefore d = \left[\frac{175.81 Q^2 (l + 4.5d)}{h}\right]^{\frac{1}{5}}. \qquad (1)$$

Or, by logarithms,

$\log d = \frac{1}{5}[2.2450532 + 2 \log Q + \log (l+4.5d) - \log h]$, (2)

where d is in inches, and all the other terms are in feet.

When the pipes are very long, or when d is small as compared with l, (2) becomes

$\log d = \frac{1}{5}(2.2450532 + 2 \log Q + \log l - \log h)$. (3)

REM.—The value of d can be obtained from (1) only by successive approximations. When considerable accuracy is required, find the value of d from (3), and substitute it in (2), which will give a first approximate value of d; and this again substituted in (2) will give a closer approximate value; and so on to any required degree of accuracy. Generally the first approximate value will be found sufficiently accurate for all practical purposes.

EXAMPLE.

What is the diameter of a pipe which shall deliver 25000 gallons of water per hour, when the length of the pipe is 2500 feet, and the head of water 225 feet?

Here $h = 225$ and $l = 2500$, and the number of cubic feet delivered per second is

$$Q = \frac{25000}{60 \times 60 \times 6.2322} = 1.1145 \text{ nearly.}$$

Substituting in (3), we have

$\log d = \frac{1}{5}(2.2450532 + 2 \log 1.1145 + \log 2500 - \log 225)$

$\qquad = \frac{1}{5}(\overline{1}.9870309 + 3.3979400) = .67699$;

$\therefore d = 4.7533$ inches.

Substituting this value of d in (2), we have

$$\log d = \tfrac{1}{3}[1.9870309 + \log (2500 + 4.5 \times 4.7533)]$$
$$= .6777337;$$
$$\therefore d = 4.761 \text{ inches},$$

which approximation is sufficiently accurate for all practical purposes.

108. Sudden Enlargement of Section.—Whenever there is a change in the cross-section of a pipe or any other conduit, there is a change of velocity, the velocity being inversely proportional to the cross-section of the stream (Art. 75). If the cross-section of a pipe is suddenly changed, there is a sudden change in the velocity of the current of water, and therefore there is a loss of kinetic energy. Thus, suppose the pipe AECF is suddenly enlarged in section at BD; then, as the water in the smaller pipe has a greater velocity than the water in the larger one, there will be an abrupt change of velocity at BD, and this change of velocity will be accompanied by a loss of kinetic energy, in the same way as when two inelastic bodies impinge upon each other.

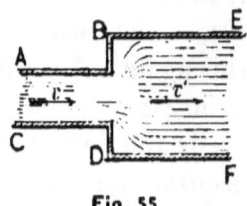

Fig. 55

Let v and v' be the velocities of the water in the smaller and larger pipes, respectively, a and a' the areas of the sections of these pipes, and W and W'' the weights of water discharged from them per second.

Now as the water moves out of the smaller pipe into the larger one, it impinges against the more slowly moving current in that pipe, and after the impact the two bodies of water, W and W'', move on together with the common velocity v'. And since in this case W is very small compared with W'', we have, from (3) of Art. 98, work lost by the water at the abrupt change of velocity

$$= (v - v')^2 \frac{W}{2g}. \qquad (1)$$

If h_1 is the head of water corresponding to this loss of work, we have, for the work lost,

$$h_1 W = (v - v')^2 \frac{W}{2g}; \qquad (2)$$

and therefore the head lost is

$$h_1 = \frac{(v - v')^2}{2g}. \qquad (3)$$

Hence, *the head lost at the abrupt change of velocity is measured by the height due to this change of velocity.*

Since we have $\quad v : v' :: a' : a,$

$$\therefore v = \frac{a'}{a} v';$$

substituting this in (3), we have, for the loss of head,

$$h_1 = \left(\frac{a'}{a} - 1\right)^2 \frac{v'^2}{2g} = \beta \frac{v'^2}{2g}, \qquad (4)$$

where we put $\quad \beta = \left(\frac{a'}{a} - 1\right)^2, \qquad (5)$

which is the corresponding *coefficient of resistance*.*

SCH.—When the edges are rounded off so as to cause a gradual passage from one pipe into the other, and the difference in the pipes is small, the loss of work as shown by experiment is very small.

---

* First found by Borda. (See Weisbach's Mechs., p. 884.)

109. Sudden Contraction of Section.

When water passes from a larger to a smaller section, as in Figs. 56, 57, where it passes from the pipe AB into the narrower pipe CEDF, a contraction is formed, and the contracted stream abruptly expands to fill the section of the pipe, thereby causing a loss of head by this sudden enlargement, precisely as in Art. 108.

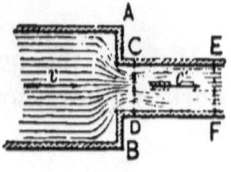

Fig. 56

(1) Let a be the area of the section and v the velocity of the stream at EF. Then, if c is the coefficient of contraction, the section of the stream at CD will be ca, and the velocity v' at this section will be found by means of the formula

$$v'ca = va;$$

$$\therefore v' = \frac{v}{c}.$$

Hence, the loss of head in passing from CD to EF is

$$h_1 = \frac{(v' - v)^2}{2g} = \left(\frac{1}{c} - 1\right)^2 \frac{v^2}{2g}$$

$$= \beta \frac{v^2}{2g}, \tag{1}$$

if β is put for $\left(\frac{1}{c} - 1\right)^2$.

If c is taken $= 0.64$ (Art. 92), we have

$$\beta = \left(\frac{1 - 0.64}{0.64}\right)^2 = 0.316,$$

which in (1) gives $h_1 = 0.316 \dfrac{v^2}{2g}.$ (2)

SCH. 1.—The value of the coefficient of contraction in this case is, however, not well ascertained, and the result is somewhat modified by friction. For water entering a cylindrical pipe from a reservoir of indefinitely large size, experiment shows that β is increased by the resistance at the entrance into the pipe, and by the friction of the water in the pipe, to 0.505, so that (1) becomes

$$h_1 = 0.505 \frac{v^2}{2g}. \tag{3}$$

(2) If there is a diaphragm at the mouth of the pipe, as at AB, Fig. 57, with an opening ab, whose cross-section is smaller than the cross-section of the pipe CEDF, let a' be the area of this orifice. Then if c is the coefficient of contraction as before, the area of the contracted stream is ca', and the velocity v' at the contracted section will be

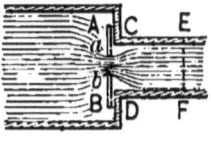

Fig. 57

$$v' = \frac{a}{ca'} v;$$

hence, the head lost in passing from the contracted section to the pipe at EF is

$$h_1 = \frac{(v' - v)^2}{2g} = \left(\frac{a}{ca'} - 1\right)^2 \frac{v^2}{2g}$$

$$= \beta \frac{v^2}{2g}, \tag{4}$$

where the corresponding *coefficient of resistance* is

$$\beta = \left(\frac{a}{ca'} - 1\right)^2. \tag{5}$$

SCH. 2.—Weisbach has found experimentally the following values of the coefficients c and β, when the stream approaching the orifice, as in Fig. 57, was considerably larger than the orifice.*

---

* Ency. Brit., Vol. XIII., Art. *Hydromechanics*.

$\frac{a'}{a} =$	0.1	0.2	0.3	0.4	0.5
$c =$	.616	.614	.612	.610	.607
$\beta =$	231.7	50.99	19.78	9.612	5.256
$\frac{a'}{a} =$	0.6	0.7	0.8	0.9	1.0
$c =$	.605	.603	.601	.598	.596
$\beta =$	3.077	1.876	1.169	0.734	0.480

SCH. 3.—When the edges are rounded off so the contraction is very gradual, the loss of work is very much diminished.

EXAMPLE.

What is the discharge through the orifice in Fig. 57, when the head is 1½ feet, the diameter of the contracted circular orifice = 1½ inches, and that of the pipe CEDF = 2 inches?

Here we have

$$\frac{a'}{a} = \left(\frac{1\frac{1}{2}}{2}\right)^2 = \frac{9}{16} = 0.56,$$

and therefore from the table $c = 0.606$, and from (5),*

$$\beta = \left(\frac{16}{9 \times 0.606} - 1\right)^2 = 3.74.$$

Hence, the whole head is

$$h = \frac{v^2}{2g} + h_1 = (1 + \beta)\frac{v^2}{2g};$$

---

* Since c comes between two numbers in the table, β is found more accurately from (5).

therefore, the velocity of efflux is

$$v = \frac{\sqrt{2gh}}{\sqrt{1+\beta}} = \frac{8.025\sqrt{1.5}}{\sqrt{4.74}} = 4.51,$$

and consequently the discharge, from (1) of Art. 106, is

$$Q = \frac{\pi}{4} d^2 v = \frac{\pi}{4} \times 4 \times 4.51 \times 12$$

$$= 170 \text{ cubic inches.}$$

110. Elbows.— When pipes are bent so as to form *elbows*, they present resistances to the motion of water in them; and these resistances, like many other phenomena of efflux, can be determined only by experiment. If a pipe ACB forms an elbow, the stream separates itself from the inner surface of the second branch of the pipe, in consequence of the centrifugal force. If the second branch is very short, terminating, for instance, at ab, the efflux will be smaller than the full cross-section of the pipe. But if the second branch is longer, terminating at B, an eddy is formed at D, and beyond this the pipe is again filled, so that the velocity of efflux v is less than the velocity at D. This diminution of the velocity of efflux must be treated in the same way as the resistance produced by a contraction in the pipe (Art. 109).

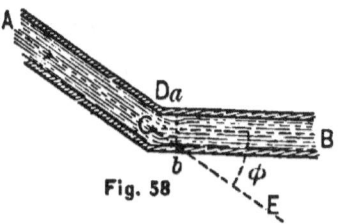
Fig. 58

Hence, if a is the cross-section of the pipe, and c is the coefficient of contraction, the section of the stream at D is ca, and the velocity v' of the contracted stream is

$$v' = \frac{v}{c},$$

and hence the loss of head in passing from D to B is

$$h_1 = \frac{(v' - v)^2}{2g} = \left(\frac{1}{c} - 1\right)^2 \frac{v^2}{2g}$$

$$= \beta \frac{v^2}{2g}. \quad (1)$$

The coefficient of contraction c, and therefore the corresponding coefficient of resistance β, depends upon the *angle of deviation* BCE. From experiments with a pipe $1\frac{1}{4}$ inches in diameter, Weisbach found the coefficient of resistance to be

$$\beta = 0.9457 \sin^2 \frac{\phi}{2} + 2.047 \sin^4 \frac{\phi}{2}, \quad (2)$$

by which he computed a series of coefficients of resistance for different angles of deviation.* From (2) it follows that the kinetic energy of water in pipes is considerably diminished by elbows. If the elbow is right-angled, we have, from (2), $\beta = 0.9846$, which in (1) gives

$$h_1 = 0.9846 \frac{v^2}{2g};$$

hence, at a right-angled elbow, very nearly the whole head due to the velocity is lost.

Sch.—If to one elbow ACB another elbow is joined, the second one turning the stream to the same side as the first one, there is no further contraction of the stream, and therefore, for efflux with full cross-section, β is no larger than for a single elbow. But if the second elbow turns the stream to the opposite side, the contraction is a double one, and the coefficient of resistance is consequently twice as great as for a single elbow.

---

* Ency. Brit., Vol. XII., p. 487.

111. Bends.—When the pipes have curved bends, the resistance is much less than in elbows. If a pipe ACB is curved, it also, in consequence of the centrifugal force, causes the stream to separate itself from the concave surface, and to form a partial contraction. If the bend terminates at BD, the cross-section of the stream at its outlet is smaller than that of the pipe. But if the bend is terminated by a long straight pipe BF, an eddy is formed at D, and beyond this the pipe is again filled, so that the velocity of efflux v is less than the velocity at D.

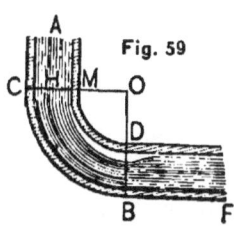

Fig. 59

If c is the coefficient of contraction, the velocity v' of the contracted stream is

$$v' = \frac{v}{c},$$

and hence the loss of head in passing from D to F :

$$h_1 = \frac{(v' - v)^2}{2g} = \left(\frac{1}{c} - 1\right)^2 \frac{v^2}{2g}$$

$$= \beta \frac{v^2}{2g}. \tag{1}$$

This is Weisbach's method, but the coefficient of contraction for bends is not very satisfactorily ascertained.

If r = the radius of the pipe = MH = HC, and ρ = the radius of curvature = HO, then Weisbach's formula for the coefficient of resistance at a bend in a pipe of circular section is

$$\beta = 0.131 + 1.847 \left(\frac{r}{\rho}\right)^{\frac{7}{2}}; \tag{2}$$

and for bends with rectangular cross-sections,

$$\beta = 0.124 + 3.104 \left(\frac{s}{2\rho}\right)^{\frac{7}{2}}, \tag{3}$$

where s is the length of the side of the section parallel to the radius of curvature ρ. (See Weisbach's Mechs., p. 897.)

111a. Pipe of Uniform Diameter Equivalent to one of Varying Diameter.

—Pipes for the supply of towns * often consist of a series of lengths, the diameter for each length being the same, but differing from those of the other lengths. In approximate calculations of the head lost in such pipes, it is generally accurate enough to neglect the smaller losses of head and to regard only the friction of the pipe, and then the calculations may be facilitated by reducing the pipe to one of uniform diameter, having the same loss of head. Such a uniform pipe is called an *equivalent pipe*.

Let A be the pipe of variable diameter, and B the equivalent pipe of uniform diameter. In A let l_1, l_2, etc., be the lengths, d_1, d_2, etc., the diameters, v_1, v_2, the velocities for the successive portions, and let l, d, v, be the corresponding quantities for the equivalent uniform pipe. Then the total loss of head in A due to friction is

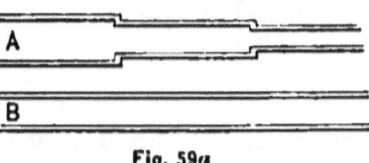

Fig. 59a

$$h = f \frac{l_1}{d_1} \frac{v_1^2}{2g} + f \frac{l_2}{d_2} \frac{v_2^2}{2g} + \text{etc.,}$$

and in the uniform pipe B,

$$h = f \frac{l}{d} \frac{v^2}{2g}.$$

If these pipes are equivalent, we have

$$f \frac{l}{d} \frac{v^2}{2g} = f \frac{l_1}{d_1} \frac{v_1^2}{2g} + f \frac{l_2}{d_2} \frac{v_2^2}{2g} + \text{etc.} \qquad (1)$$

---

* Such pipes are called *water mains*.

But since the discharge is the same for all portions,

$$\pi \frac{d^2}{4} v = \pi \frac{d_1^2}{4} v_1 = \pi \frac{d_2^2}{4} v_2 = \text{etc.}$$

$$\therefore \quad v_1 = v \frac{d^2}{d_1^2}; \quad v_2 = v \frac{d^2}{d_2^2}; \quad \text{etc.} \qquad (2)$$

Then supposing that f is constant for all the pipes, we have, from (1) and (2),

$$l = d^5 \left(\frac{l_1}{d_1^5} + \frac{l_2}{d_2^5} + \frac{l_3}{d_3^5} + \text{etc.} \right), \qquad (3)$$

which gives the length of the equivalent uniform pipe which would have the same total loss of head, for any given discharge, as the pipe of varying diameter.

Cor.—If the lengths of the successive portions are all equal, we have $l_1 = l_2 = l_3 = $ etc., and (3) becomes

$$l = l_1 d^5 \left(\frac{1}{d_1^5} + \frac{1}{d_2^5} + \frac{1}{d_3^5} + \text{etc.} \right) \qquad (4)$$

111b. Pipe of Uniform Diameter with Discharge Diminishing Uniformly along its Length.—In the case of a branch main, the water is delivered at nearly equal distances to service pipes along the route. Let AB be a main of diameter d and length L; let Q_0 cubic feet per second enter at A, and let q cubic feet per foot of its length be delivered to service pipes. Then at any point C, l feet from A, the discharge is $Q = Q_0 - ql$. Consider a short length dl at P. The loss of head in that length is

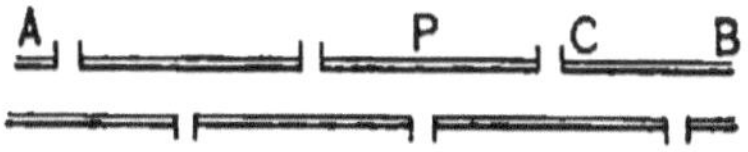

Fig. 59b

$$f \frac{dl}{d} \frac{v^2}{2g} = 16 f \cdot \frac{Q^2}{\pi^2 d^5} \frac{dl}{2g}$$

$$= \frac{8f}{\pi^2 d^5 g} (Q_0 - ql)^2 \, dl.$$

Hence, the whole head lost in the length AB is

$$h = \frac{8f}{\pi^2 d^5 g} \int_0^L (Q_0 - ql)^2 \, dl$$

$$= \frac{8fL}{\pi^2 d^5 g} (Q_0^2 - q Q_0 L + \tfrac{1}{3} q^2 L^2) ; \qquad (1)$$

or, putting $P = qL$, the total discharge through the service pipes between A and B, (1) becomes

$$h = \frac{8fL}{\pi^2 d^5 g} (Q_0^2 - P Q_0 + \tfrac{1}{3} P^2). \qquad (2)$$

The discharge at the end B of the pipe is $Q_0 - P$. If the pipe is so long that $Q_0 - P = 0$, all the water passes into the service pipes, and (2) becomes

$$h = \frac{1}{3} \frac{8fL}{\pi^2 d^5 g} P^2. \qquad (3)$$

(See Ency. Brit., Vol. XII., p. 486.)

112. General Formula when all the Resistances to the Flow of Water are Considered.—Let β_1 be the coefficient of resistance for enlargements and contractions (Arts. 108 and 109), and β_2 the coefficient of resistance for elbows and bends (Arts. 110 and 111). Then adding together (3) of Art. 105, (4) of Arts. 108 and 109, (1) of Arts. 110 and 111, we have for the entire head h,

$$h = (1 + \beta) \frac{v^2}{2g} + (a_1 v + a_2 v^2) \frac{l}{d} + \beta_1 \frac{v^2}{2g} + \beta_2 \frac{v^2}{2g}$$

$$= (a_1 v + a_2 v^2) \frac{l}{d} + (1 + \beta + \beta_1 + \beta_2) \frac{v^2}{2g}, \qquad (1)$$

where the values of a_1 and a_2 are given in Art. 105, β_1 in Art. 109, β_2 in Art. 110, and $\beta = .505$ to $.08$.

Neglecting a_1, since it is very small compared with a_2 (Art. 105), and putting $f = 2ga_2 = .03$, from (1) of Art. 105, we have, from (1),

$$h = \left(f\frac{l}{d} + 1 + \beta + \beta_1 + \beta_2\right)\frac{v^2}{2g}. \qquad (2)$$

SCH.—An enlargement should be made in the pipe at any considerable bendings; and when any change takes place in the diameter of the pipe, the parts at the junction should be rounded off. At all considerable bends, where the pipe changes from ascending to descending, a provision should be made for clearing the pipe of the air which is disengaged from the water. Unless some provision is made for the escape of this air, it will accumulate in the highest bends and obstruct the flow of the current.

EXAMPLE.

In the example of Art. 105 there are 40 bends in the pipe, each having the radius of curvature exceeding ten times the radius of the pipe. Find the velocity of efflux.

Here $\quad f = .03, \quad \beta = .505, \quad \beta_1 = 0,$

$$\beta_2 = .131 + 1.847\left(\tfrac{1}{10}\right)^{\frac{7}{2}} = .1312;$$

$$\therefore \; 40\beta_2 = 5.248, \quad l = 5780, \quad d = .5, \quad h = 170.$$

Hence, from (2), we have

$$170 = \left(.03 \times \frac{5780}{.5} + 1.505 + 5.248\right)\frac{v^2}{64\tfrac{1}{3}}$$

$$= 353.553\,\frac{v^2}{64\tfrac{1}{3}};$$

$$\therefore \; v = 5.562 \text{ feet per second.}$$

For practical calculations on the flow of water in pipes, see Ency. Brit., Vol. XII., p. 488.

113. Flow of Water in Rivers and Canals.—

When water flows in a pipe, the section at any point is determined by the form of the boundary. When it flows in an open channel with free upper surface, the section depends on the velocity due to the kinetic conditions. The bottom of the channel and the two banks are called *the bed of the stream*. A section of the stream at right angles to the direction in which it is flowing is called a *transverse section*, and of the line bounding this section, the part that is beneath the water surface is called the *wetted perimeter*. A vertical section in the direction of the stream is called the *longitudinal section* or *profile*.

Let ABCD represent a longitudinal section of a limited portion of a stream, AD, BC, two transverse sections, AB the surface of the stream, DC the bottom of the channel, and AE a horizontal line. Let $l = $ the length of AB in feet; $h = $ BE, the difference of level of the water surface at the two extremities of the distance l; $\theta = $ the angle BAE, the slope of the stream: $\sin \theta = \dfrac{h}{l} = $ the sine of the slope, or the fall of the water surface in one foot; $a = $ the area of the transverse section at BC in square feet; $p = $ the length of the wetted perimeter of the transverse section at BC; $r = \dfrac{a}{p}$, the hydraulic mean depth, or the mean radius of the section; $Q = $ the discharge through the section at BC in cubic feet per second; $v = \dfrac{Q}{a} = $ the mean velocity of the stream in feet per second, which is taken as the common velocity of all the particles.

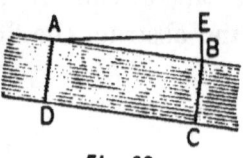

Fig. 60

114. Different Velocities in a Cross-Section.—The velocity of the water is not uniform in all points of the same transverse section. In all actual streams the different fluid filaments have different velocities. The adhesion of the water to the bed of the channel, and the cohesion of the molecules of water cause the particles of water nearest to the sides and bed of the channel to be most hindered in their motion. For this reason, the velocity is much less at the bottom and sides than it is at the surface and centre. According to some authors, the maximum velocity in a straight river is generally found in the middle of its surface, or in that part of the surface where the water is the deepest.* Theoretically we should expect this, but practically it is often very different.

The theory adopted by most modern writers is the following: The motion of the water being caused solely by the slope of the surface, the velocity in all parts of any transverse section of the river would be equal, were it not for the retarding influence of the bed. The layer of elementary particles next to the bed adheres firmly to it by virtue of the force of adhesion. The next layer is retarded partly by the cohesion existing between it and the first, partly by the friction, and partly by the loss of kinetic energy arising from constant collision with the irregularities which correspond to those of the bed. The next layer is retarded in the same manner, but in a less degree. Thus, according to this theory, the effect of the resistances is diminished as the distance from the bed is increased; and assuming, as is usually done, that no sensible resistance is experienced from the air, the maximum velocity should be found in the surface filament situated at the greatest distance from the bed. The many experiments, however, which have been made to determine the actual variation in velocity at different depths, and upon the surface, at different distances from the banks, give very different results.

---
* Weisbach's Mechs., p. 956; also Tate's Mech. Phil., p. 302.

Focacci found that in a canal 5 feet deep, the maximum velocity was from 2 to 2.5 feet below the surface.

Defontaine states that in calm weather the velocity of the Rhine is greatest at the surface.

Raucourt made experiments upon the Neva where it is 900 feet wide and of regular section, the maximum depth being 63 feet. When the river was frozen over, the maximum velocity (2 feet 7 inches per second) was found a little below the middle of the deepest vertical, where it was nearly double the velocity at the surface and bottom, which were nearly equal to each other. In summer, he found the maximum velocity was near the surface in calm weather; but when a strong wind was blowing up stream, the surface velocity was greatly diminished, so that it hardly exceeded that at the bottom. He considers the law of diminution of velocity to be given by the ordinates of an ellipse whose vertex is a little below the bottom, and whose minor axis is a little below the surface.

Hennocque found the maximum velocity in the Rhine to be, in calm weather, or with a light wind, $\frac{1}{4}$ of the depth below the surface; in a strong wind up stream, it was a little below mid-depth; in a strong wind down stream, it was at the surface.

Baumgarten found in the Garonne that the maximum velocity was generally at the surface, but that in one section (about 325 feet wide) it was always below the surface.

D'Aubuisson considers that the velocity diminishes slowly at first, as the depth increases, but that near the bottom it is more rapid. The bottom velocity, however, is always more than half that of the surface.

Boileau found, by experiment in a small canal, that the maximum velocity was $\frac{1}{4}$ to $\frac{1}{5}$ of the depth below the surface. Below this point, the velocity diminished rapidly, and nearly in the ratio of the ordinates of the parabola whose axis was at the surface. He decided, from a discussion of the experiments of Defontaine, Hennocque, and Baumgarten, that in large rivers the maximum velocity is by no means always at the surface.*

It will be seen from this synopsis that there is a great diversity among the results obtained by different experimenters, and that no mathematical relation, of sufficiently general application to constitute a practical law, has been yet discovered.

---

* See Report on the Hydraulics of the Mississippi River, by Humphreys and Abbott, pp. 200, etc.

214   DIFFERENT VELOCITIES IN A CROSS-SECTION.

The velocities observed on any given longitudinal section, at any given moment, do not form, when plotted, any regular curve. But if a series of observations are taken at each depth, and the results averaged, the mean velocities at each depth, when plotted, give a regular curve agreeing very fairly with a parabola whose axis is horizontal, corresponding to the position of the filament of maximum velocity. All the best observations show that the maximum velocity is to be found at some distance below the free surface.

In the experiments on the Mississippi River, the velocities on any longitudinal section, in calm weather, were found to be represented very fairly by a parabola, the greatest velocity being at $\frac{3}{10}$ of the depth of the stream from the surface. With a wind blowing down stream, the surface velocity is increased and the axis of the parabola approaches the surface. With a wind blowing up stream, the surface velocity is diminished and the axis of the parabola is lowered, sometimes to half the depth of the stream. The observers on the Mississippi drew from their observations the conclusion that there was an energetic retarding action at the surface of a stream, like that at the bottom and sides. If there were such a retarding action, the position of the filament of maximum velocity below the surface would be explained. If there were no such resistance, the maximum velocity should be at the surface.

It is not difficult to understand that a wind, acting on surface ripples, should accelerate or retard the surface motion of the stream, and the Mississippi results may be accepted so far as showing that the surface velocity of a stream is variable when the mean velocity of the stream is constant. Hence observations of surface velocity, by floats or otherwise, should only be made in very calm weather. But it is very difficult to suppose that, in still air, there is a resistance at the free surface of the stream at all analogous to that at the sides and bottom. In very careful experiments, Boileau found the maximum velocity, though raised a little above its position for calm weather, still at a considerable distance below the surface, even when the wind was blowing down stream with a velocity greater than that of the stream, and when the action of the air must have been an accelerating and not a retarding action. Prof. James Thomson has given a much more probable explanation of the diminution of the velocity at and near the free surface. He points out that portions of water, with a diminished velocity from retardation by the sides or bottom, are thrown off in eddying masses and mingle with the rest of the stream. These eddying masses modify the velocity in all parts of the stream, but have their greatest influence at the free surface. Reaching the free surface, they spread out and remain there, mingling with the water at that level, and diminishing the velocity which would otherwise be found there.*

115. Transverse Section of the Stream. — The form of the transverse section and the direction of the current have such an effect upon the velocity at the surface, at different distances from the banks, that there can be no definite law of change. There is generally an increase of velocity, as the distance from the banks is increased, until the maximum point is reached. That portion of the river where the water has its maximum velocity is called the *line of current* or *axis of the stream*, and the deepest portion of the stream is called the *mid-channel*. When the stream bends, its axis is generally near the concave shore.

It is observed that the surface of a stream, in any cross-section, is highest where the velocity is greatest, which is accounted for by the fact that, when the water is in motion, it exerts less pressure at right angles to the direction of its motion than when it is at rest, and therefore, where the velocity is greatest the water must be highest, to balance the pressure at the sides, where the velocity is less.

It frequently happens that, while the mass of the water in a river is flowing on down the river, the water next the shore is running up the river. It is no unusual thing to find a swift current and a corresponding fall on one shore *down stream*, and on the opposite shore a visible current and an appreciable fall *up stream*; *i. e.*, on one side of the river the water is often running rapidly up stream, while on the other side it is running with equal or greater rapidity down stream. The apparent slope at every point is affected by the bends of the river, and by the centrifugal force acquired by the water in sweeping round the curves, and by the eddies which form on the opposite side. The surface of the river is not therefore *a plane*, but a complicated *warped*

---

* Ency. Brit., Vol. XII., p. 497.

surface, varying from point to point, and inclining alternately from side to side.*

116. Mean Velocity.—The mean velocity of the water in a cross-section is equal to the quotient arising from dividing the discharge per second by the area of the transverse section.

When the discharge per second is not known, the mean velocity may be determined by measuring the velocities in all parts of the transverse section, and taking a mean of the results. If the transverse section is irregular in form, the only accurate manner of determining the mean velocity is to divide this section into partial areas so small that the velocity throughout each may be considered invariable. The discharge is then equal to the sum of the products of these partial areas by their velocities.

Let a_1, a_2, a_3, etc., be the small partial areas into which the transverse section is divided, and v_1, v_2, v_3, etc., the velocities in these small areas. Then the whole area is

$$a = a_1 + a_2 + a_3 + \text{etc.,} \quad (1)$$

and the whole discharge is

$$av = a_1 v_1 + a_2 v_2 + a_3 v_3 + \text{etc.;} \quad (2)$$

therefore the mean velocity is

$$v = \frac{a_1 v_1 + a_2 v_2 + a_3 v_3 + \text{etc.}}{a_1 + a_2 + a_3 + \text{etc.}} \quad (3)$$

117. Ratio of Mean to Greatest Surface Velocity. —It is often very important to be able to deduce the mean velocity from observation of the greatest surface velocity. The greatest surface velocity may be determined by floats. Unfortunately, however, the ratio of the maximum surface velocity to the mean velocity is extremely variable; and it

---
* See Report on the Mississippi.

has formed the subject of much careful investigation. Putting v_0 for the greatest surface velocity, and v_m for the mean velocity of the whole cross-section, the following values have been found for $\dfrac{v_m}{v_0}$:

De Prony, experiments on small wooden channels, 0.8164
Experiments on the Seine, ... 0.62
Destrem and De Prony, experiments on the Neva, 0.78
Boileau, experiments on canals, ... 0.82
Baumgarten, experiments on the Garonne, . . . 0.80
Brünings (mean), ... 0.85
Cunningham, Solani aqueduct, ... 0.823
Dubuat, experiments on small canals (mean), . . 0.83
Dupuit, from theoretical considerations, believes the ratio to vary between 0.67 and 1.00.

Various formulæ have been proposed for determining the ratio $\dfrac{v_m}{v_0}$. Bazin found from his experiments the following empirical expression,

$$v_m = v_0 - 25.4\sqrt{r\theta}, \qquad (1)$$

where r is the hydraulic mean depth, and θ the slope of the stream (Art. 113).

Prony found the following formula.

$$v_m = \frac{v_0(v_0 + 7.77)}{v_0 + 10.33}. \qquad (2)$$

The ratio of the mean velocity to the surface velocity in one longitudinal section is better ascertained than the ratio of the greatest surface velocity to the mean velocity of the whole cross-section. Let the river be divided into a number of compartments by equidistant longitudinal planes, and the surface velocity be observed in each compartment; then from this the mean velocity in each compartment and the discharge can be computed. The sum of the partial discharges will be the total discharge of the stream. The following formula* is convenient for determining the ratio of the surface velocity to the mean velocity in the same vertical. Let v be the mean and V the surface velocity in any given vertical longitudinal section, the depth of which is h.

$$\frac{v}{V} = \frac{1 + 0.1478\sqrt{h}}{1 + 0.2216\sqrt{h}}. \qquad (3)$$

Sch.—In the gaugings of the Mississippi, it was found that the mid-depth velocity differed by only a very small quantity from the mean velocity in the vertical section, and it was uninfluenced by wind. If therefore a series of mid-depth velocities are determined, they may be taken to be the mean velocities of the compartments in which they occur, and no formula of reduction is necessary.

118. Processes for Gauging Streams.—The discharge of large creeks, canals, and rivers, can be measured only by means of *hydrometers*, which are instruments for indicating the velocity. The simplest of these instruments are *surface floats*; these are convenient for determining the surface velocities of a stream, though their use is difficult near the banks. Any floating body can be used for this purpose; but it is safer to employ bodies of medium size, and of but little less specific gravity than the water itself. Very large bodies do not easily assume the velocity of the water, and very small bodies, especially when they project much above the surface of the water, are easily disturbed in their motion by accidental circumstances, such as wind, etc.

The floats may be small balls of wood, of wax, or of hollow metal, so loaded as to float nearly flush with the water surface. To make them visible, they may have a vertical painted stem. In experiments on the Seine, cork balls 1¾

---

* Given by Eyner in Erbkam's Zeitschrift for 1875.

inches diameter were used, loaded to float flush with the water surface, and provided with a stem. Bits of solid wood, and bottles filled with water until nearly submerged, have often been used for surface floats. Boileau proposes balls of soft wax, on account of their adhesive properties. In Captain Cunningham's observations, the floats were thin circular disks of English deal, 3 inches diameter and $\frac{1}{4}$ inch thick. For observations near the banks, floats 1 inch diameter and $\frac{1}{8}$ inch thick were used. To render them visible, a tuft of cotton wool was used, loosely fixed in a hole at the centre.

The velocity is obtained by allowing the float to be carried down, and noting the time of passage over a measured length of the stream. If t is the time in which the float passes over a length l, which has been previously measured, and staked off on the shore, then the velocity v is $v = \frac{l}{t}$.

To mark out distinctly the length of stream over which the floats pass, two ropes may be stretched across the stream at a distance apart, which varies usually from 50 to 250 feet, according to the size and rapidity of the river. To mark the precise position at which the floats cross the ropes, Capt. Cunningham, in his experiments, used short white rope pendants, hanging so as nearly to touch the water. In this case the streams were 80 to 180 feet wide.

In wider streams the use of ropes to mark the length of run is impossible; in such cases, recourse must be had to some such method as the following: Let AB be the measured length $= l$, on one side of the river. Put two rods C and D, by means of a suitable instrument, in such a position upon the other side of the river that the lines CA and DB shall be perpendicular to AB. Then the observer, placed behind A, notes by his watch the instant the float E, which has been placed in the water some distance above, arrives at the line

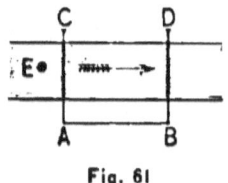

Fig. 61

AC, and then, passing down to B, he observes the instant that the float arrives at the line BD. By subtracting the time of the first observation from that of the second, he obtains the time t in which the space l is described.

For measuring the velocity below the surface, double floats* are used. They are of various kinds, usually consisting of small surface floats, supporting by cords larger submerged bodies. Suppose two equal and similar floats, connected by a string, wire, or thin wire chain. Let one float be a little heavier, and the other a little lighter than water, so that only a small portion of the latter will project above the surface of the water. We first determine by a single float the surface velocity v_s; we then determine the velocity of the connected floats, which will be the mean of the surface velocity and the velocity at the depth at which the heavier float swims. If v_d is the velocity at the depth to which the lower float sinks, we have, calling v the mean velocity,

$$v = \frac{v_s + v_d}{2};$$

$$\therefore \; v_d = 2v - v_s. \qquad (1)$$

By connecting the floats successively by longer and longer pieces of wire, we obtain in this way the velocities at greater and greater depths.

To obtain the mean velocity in a perpendicular, a floating staff or rod is often employed. This consists of a cylindrical rod, loaded at the lower end so as to float nearly vertical in water. A wooden rod, with a metal cap at the bottom in which shot can be placed, so as to prevent more than the head from projecting above the water surface, answers well, and sometimes the wooden rod is made of short pieces which can be screwed together so as to suit streams of different depths. A tuft of cotton wool at the top serves to make

---

* First used by da Vinci.

the float more easily visible. Such a rod, so adjusted in length that it sinks nearly to the bed of the stream, gives directly the mean velocity of the whole vertical section in which it floats. (For a complete description of gauging streams, see "Report on the Mississippi.")

119. Most Economical Form of Transverse Section.—The best form of the transverse section must be that which presents the least resistance to a given quantity of water flowing through the channel. From Art. 103, the resistance of the bed of the stream, in consequence of the adhesion and friction, varies directly as the surface of contact, and consequently as the wetted perimeter p (Art. 113), and inversely as the area of the transverse section, $i.\,e.$, the resistance of the bed of the stream varies as $\frac{p}{a}$. In order, therefore, to have the least resistance from friction, the form of the section must be that which has the least perimeter for a given area, $i.\,e.$, the wetted perimeter p must be a minimum for a given area a, or the area must be a maximum for a given wetted perimeter. Now, among all figures of the same number of sides, the regular one, and among all the regular ones, the one with the greatest number of sides has the smallest perimeter for a given area. Hence, for closed pipes, the resistance of friction is the smallest when the transverse section is a circle; but in open channels, the upper surface, being free, or in contact with the air alone, must not be included in the perimeter.

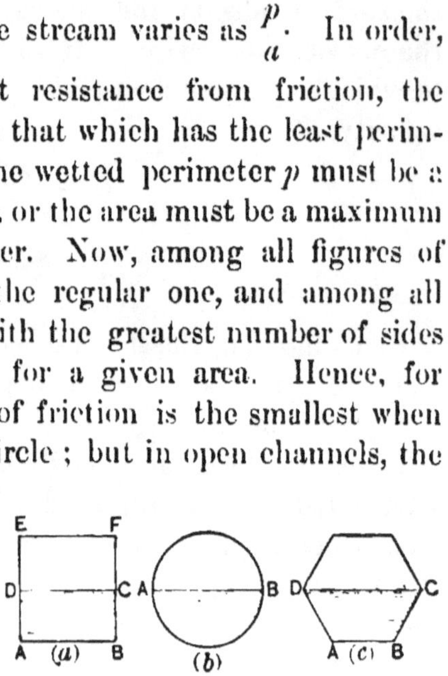

A horizontal line DC, passing through the centre of the square AF, divides

Fig. 62

the area and perimeter into two equal parts, and what has been said of the square is true of these halves; hence, of all rectangular forms of transverse sections, the half square

ABCD is the one which causes the least resistance of friction, and therefore is the best for open channels. Also, of all trapezoidal sections, the semi-hexagon ABCD is the one which causes the least resistance of friction; and so on to the other cases. But the semicircle will present less resistance of friction than the semi-hexagon, and this latter less than the semi-square. The half decagon offers still less resistance than the half hexagon or the half square. The circular and square sections are used only for troughs made of iron, stone, or wood. The trapezoid is employed in canals, which are dug out or walled up. It is very rare that other forms are used, owing to the difficulty of constructing them.

120. Trapezoidal Section of a Canal of Least Resistance, when the Slope of the Sides is Given.
—Let ABCD be the section. Put
$x =$ AB, the width of the bottom,
$y =$ BE, the depth, and $\theta =$ BCE,
the angle of the slope, which is to
be considered as a given quantity,
dependent upon the nature of the
ground in which the canal is excavated, and $a =$ the given area of the section ABCD.

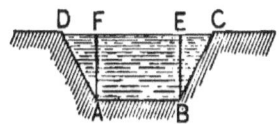

Fig. 63

Then the wetted perimeter of the section is

$$p = AB + 2BC$$
$$= x + 2y \operatorname{cosec} \theta. \qquad (1)$$

The area of the section is

$$a = xy + y^2 \cot \theta;$$
$$\therefore x = \frac{a - y^2 \cot \theta}{y}, \qquad (2)$$

which in (1) gives

$$p = \frac{a - y^2 \cot \theta}{y} + 2y \operatorname{cosec} \theta. \qquad (3)$$

To find the value of y which makes this a minimum, we must equate to zero its derivative with respect to y, which gives

$$y^2 = \frac{a \sin \theta}{2 - \cos \theta}; \qquad (4)$$

$$\therefore y = \sqrt{\frac{a \sin \theta}{2 - \cos \theta}}. \qquad (5)$$

Hence, for a given angle of slope θ, and for a given area a, the *trapezoidal section of least resistance* is determined by (2) and (5).

Consequently, the width CD of the top is

$$CD = x + 2y \cot \theta$$

$$= \frac{a}{y} + y \cot \theta; \qquad (6)$$

and the value of $\frac{p}{a}$, from (3), is

$$\frac{p}{a} = \frac{1}{y} + \frac{2 - \cos \theta}{a \sin \theta} y$$

$$= \frac{2}{y} \quad [\text{from (4)}]. \qquad (7)$$

EXAMPLE.

What dimensions should be given to the transverse section of a canal, when the angle of slope of its banks is to be 40°, and when it is to carry 75 cubic feet of water with a mean velocity of 3 feet?

Here we have

$$a = \tfrac{75}{3} = 25 \text{ square feet};$$

and hence, from (5), we have the depth

$$y = \sqrt{\frac{25 \sin 40°}{2 - \cos 40°}} = 5\sqrt{\frac{0.64279}{1.23396}} = 3.609 \text{ feet.}$$

From (2), we have the width at the bottom,

$$x = \frac{25}{3.609} - 3.609 \cot 40°$$

$$= 6.927 - 4.301 = 2.626 \text{ feet.}$$

The width of the top, from (6), is

$$CD = 2.626 + 7.218 \cot 40° = 11.228 \text{ feet.}$$

The wetted perimeter is

$$p = x + \frac{2y}{\sin \theta}$$

$$= 2.626 + \frac{7.218}{\sin 40°} = 13.855 \text{ feet};$$

and the ratio which determines the resistance of friction is

$$\frac{p}{a} = \frac{2}{y} = 0.5542.$$

Rem.—In a transverse section in the shape of the half of a regular hexagon, where $\theta = 60°$, $x = 4.39$, $y = 3.80$, width $CD = 8.78$, and $p = 13.16$ feet, we have, for the resistance of friction,

$$\frac{p}{a} = \frac{13.16}{25} = 0.526,$$

which is less than that found for the above trapezoid.

121. Uniform Motion.—When water flows in an open channel, the velocity continues to increase so long as the accelerating force exceeds the resisting force of friction; but when these forces are equal to each other, the velocity of the stream becomes uniform. When the velocity is uniform, the entire head h is employed in overcoming the friction upon the bed. Therefore, the height of the column of water due to the resistance of friction must be equal to the fall. The height due to the resistance of friction increases with $\frac{p}{a}$, with the length l, and with the square of the velocity v (Art. 102). Hence, from (1) of Art. 103, we have

$$h = f\frac{lp}{a}\frac{v^2}{2g}, \qquad (1)$$

in which f is an empirical number, which is called *the coefficient of friction*.

Solving (1) for v, we have

$$v = \sqrt{\frac{2gha}{flp}}. \qquad (2)$$

According to Eytelwein's reduction of the ninety-one observations and experiments made by du Buat, Brünings, Funk, and Woltmann, $f = 0.007565$, which in (1) gives

$$h = 0.007565\frac{lp}{a}\cdot\frac{v^2}{2g}. \qquad (3)$$

If we put $g = 32.2$ feet, (2) and (3) become

$$v = 92.26\sqrt{\frac{ah}{pl}}, \qquad (4)$$

$$h = 0.00011747\frac{lp}{a}v^2. \qquad (5)$$

For the number of cubic feet of water flowing through the channel per second, we have

$$Q = av = 92.26a\sqrt{\frac{ah}{pl}}. \qquad (6)$$

Cor.—For pipes, we have

$$\frac{lp}{a} = \frac{l\pi d}{\frac{1}{4}\pi d^2} = \frac{4l}{d},$$

which in (3) gives

$$h = 0.03026 \frac{l}{d} \cdot \frac{v^2}{2g}, \qquad (7)$$

which agrees with (5) of Art. 104 and (1) of Art. 105.

EXAMPLE.

How much fall must a canal, whose length is 2600 feet, whose lower width is 3 feet, whose upper width is 7 feet, and whose depth is 3 feet, have in order to carry 40 cubic feet of water per second? Here we have

$$p = 3 + 2\sqrt{2^2 + 3^2} = 10.211,$$

$$a = \frac{(7+3)\,3}{2} = 15,$$

and $\qquad v = \dfrac{40}{15} = \dfrac{8}{3}.$

Substituting in (5), we have

$$h = 0.00011747 \, \frac{2600 \times 10.211}{15} \left(\frac{8}{3}\right)^2$$

$$= \frac{0.305422 \times 10.211 \times 64}{15 \times 9} = 1.48 \text{ feet.}$$

122. Coefficients of Friction.—The coefficient of friction f varies greatly with the degree of roughness of the channel sides, and somewhat also with the velocity, as in the case of pipes, increasing slightly when the velocity diminishes, and decreasing when the velocity increases. A common mean value assumed for f is 0.007565, which we used in the last Art., though it has quite a range of values. Weisbach, from 255 experiments, obtained for f the following values at different velocities:

$v =$	0.3	0.4	0.5	0.6	0.7
$f =$	0.01215	0.01097	0.01025	0.00978	0.00944
$v =$	0.8	0.9	1.0	$1\tfrac{1}{2}$	2
$f =$	0.00918	0.00899	0.00883	0.00836	0.00812
$v =$	3	5	7	10	15
$f =$	0.00788	0.00769	0.00761	0.00755	0.00750

In using this table for the value of f when v is not known, we must proceed by approximation. Determine v approximately from (4) of Art. 121. Then from this value of v find f by means of the table, and substitute the value of f so found in (2), and determine a new value of v.

EXAMPLE.

What must be the fall of a canal 1500 feet long, whose lower width is 2 feet, upper width 8 feet, and depth 4 feet, when it is required to convey 70 cubic feet of water per second?

Here we have

$$p = 2 + 2\sqrt{16 + 9} = 12,$$
$$a = 5 \times 4 = 20,$$
$$v = \tfrac{70}{20} = 3.5;$$

hence, from the table,

$$f = 0.00784.$$

Substituting in (1) of Art. 121, we have

$$h = 0.00784 \frac{1500 \times 12}{20} \times \frac{3.5^2}{2g}$$
$$= \frac{86.436}{64.4} = 1.34 \text{ feet.}$$

123. Variable Motion.—In every stream in which the discharge is constant for a given time, the velocity at different places depends on the slope of the bed. In general, the velocity will be greater as the slope of the bed is greater; and, as the velocity varies inversely as the transverse section of the stream, the section will be least where the velocity and slope are greatest. In a stream in which the velocity is variable, the work due to the fall of the stream for a given distance is equal to the work destroyed by friction together with the kinetic energy corresponding to the change of velocity, *i. e.*, the whole fall is the sum of that expended in overcoming friction, and of that expended in increasing the velocity, when the velocity increases, or if the velocity decreases, the head is the difference of these quantities.*

The resistance of friction upon a small portion of the length of the stream may be regarded as constant and measured by a head of water

$$= f \frac{lp}{a} \frac{v^2}{2g}. \qquad (1)$$

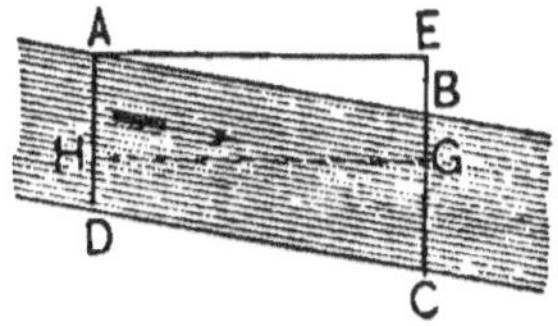

Let ABCD represent a longitudinal section of a short portion of a stream, AB the surface of the stream, and AE and HG two horizontal lines. Let $l =$ the length of AB in feet; $h =$ BE, the fall from A to B; $v_0 =$ the velocity of the stream at the upper section AD; and $v_1 =$ the velocity at the lower section BC.

Fig. 64

Now the velocity of any particle B, at the surface of the stream, is due to the height h, together with the velocity at A; hence we have, for its velocity v_1,

$$\frac{v_1^2}{2g} = h + \frac{v_0^2}{2g}. \qquad (2)$$

---

* In long rivers, with slopes not greater than 3 feet per mile, the velocity head is usually insignificant compared with the friction head. (See Fanning's Water-Supply Engineering, p. 303.)

Any particle G, beneath the surface of the water, is pressed forward by the head AH = EG, and pressed backward by the head BG; hence the head which produces motion is EG — BG = EB, or h, as before; and therefore (2) is true for any particle. Solving (2) for h, and adding the resistance of friction, as given in (1), we have

$$h = \frac{v_1^2 - v_0^2}{2g} + f\frac{lp}{a}\frac{v^2}{2g}, \tag{3}$$

in which p, a, and v denote the mean values of the wetted perimeter, the transverse section, and the velocity, respectively.

If a_0 and a_1 denote the areas of the upper and lower transverse sections, respectively, and Q the quantity of water which flows through any section in a unit of time, we have

$$a = \frac{a_0 + a_1}{2}, \tag{4}$$

and $$Q = a_0 v_0 = a_1 v_1. \tag{5}$$

From (5), we have

$$\frac{v_1^2 - v_0^2}{2g} = \left(\frac{1}{a_1^2} - \frac{1}{a_0^2}\right)\frac{Q^2}{2g}. \tag{6}$$

Now if the water flowed with the velocity v_0, we would have the head due to the resistance of friction, from (1),

$$= f\frac{lp}{a}\frac{v_0^2}{2g}; \tag{7}$$

and if it flowed with the velocity v_1, we would have the head due to the resistance of friction

$$= f\frac{lp}{a}\frac{v_1^2}{2g}. \tag{8}$$

But the former expression is less, and the latter is greater than the true head due to the resistance of friction; hence, the mean of these results will give the friction head approximately. Therefore, taking the mean of (7) and (8), and substituting for a its value from (4), we have, for the resistance of friction,

$$f\frac{lp}{a}\frac{v^2}{2g} = f\frac{lp}{a_0 + a_1}(v_0^2 + v_1^2)\frac{1}{2g}$$

$$= f\frac{lp}{a_0 + a_1}\left(\frac{1}{a_0^2} + \frac{1}{a_1^2}\right)\frac{Q^2}{2g} \qquad (9)$$

[from (5)].

Substituting (6) and (9) in (3), we have, for the whole head,

$$h = \left[\frac{1}{a_1^2} - \frac{1}{a_0^2} + f\cdot\frac{lp}{a_0 + a_1}\left(\frac{1}{a_0^2} + \frac{1}{a_1^2}\right)\right]\frac{Q^2}{2g}. \qquad (10)$$

Solving (10) for Q, we have

$$Q = \frac{\sqrt{2gh}}{\sqrt{\frac{1}{a_1^2} - \frac{1}{a_0^2} + f\frac{lp}{a_0 + a_1}\left(\frac{1}{a_0^2} + \frac{1}{a_1^2}\right)}}. \qquad (11)$$

In a prismoidal channel it will be a sufficiently close approximation to the truth to assume that the surface line of the water is straight, and then from this assumption to compute the transverse sections and their perimeters. When we have these, with the quantity of water carried and the length of a portion of the river or canal, we may determine the corresponding fall h by (10); and when we have the length, fall, and cross-section, we may determine the quantity Q by means of (11). Where greater accuracy is required, we should calculate h or Q for several small portions of the stream, and then take the arithmetic mean of the results. If only the total fall is known, this value should be substituted for h in (11), and instead of $\frac{1}{a_1^2} - \frac{1}{a_0^2}$ we should use $\dfrac{1}{a_n^2} - \dfrac{1}{a_0^2}$, where a_n denotes the area of the lowest transverse section, and instead of

$$f\frac{lp}{a_0 + a_1}\left(\frac{1}{a_0^2} + \frac{1}{a_1^2}\right),$$

the sum of all the similar values for the different portions of the stream should be used. (See Weisbach's Mechs., p. 969; also Tate's Mech. Phil., p. 305.)

EXAMPLE.

A stream falls 9.6 inches in 300 feet, the mean value of its wetted perimeter is 40 feet, the area of its upper transverse section is 70 square feet, and that of its lower is 60 square feet. Find the discharge of this stream.

From (11), we have

$$Q = \frac{8.025\sqrt{0.8}}{\sqrt{\dfrac{1}{60^2} - \dfrac{1}{70^2} + 0.007565\dfrac{300 \times 40}{130}\left(\dfrac{1}{60^2} + \dfrac{1}{70^2}\right)}}$$

$$= \frac{7.178}{\sqrt{0.0000731 + 0.0003365}} = 354\tfrac{1}{2} \text{ cubic feet.}$$

The mean velocity is

$$\frac{2Q}{a_0 + a_1} = \frac{709}{130} = 5.45 \text{ feet};$$

hence (Art. 122), a more accurate value of f is 0.00768, and therefore we have

$$Q = \frac{7.178}{\sqrt{0.0000731 + 0.0003416}} = 352.5.$$

If the same stream has at another place a fall of 11 inches in 450 feet, and if the area of its upper transverse section is

50, and that of its lower 60 square feet, the mean value of its wetted perimeter being 36 feet, we have

$$Q = \frac{8.025\sqrt{0.9167}}{\sqrt{\frac{1}{60^2} - \frac{1}{50^2} + 0.00768 \frac{450 \times 36}{110}\left(\frac{1}{60^2} + \frac{1}{50^2}\right)}}$$

$$= 8.025\sqrt{\frac{0.9167}{-0.0001222 + 0.0007549}}$$

$$= 305\tfrac{1}{2} \text{ cubic feet.}$$

The mean of these values is

$$Q = \frac{352.5 + 305.5}{2} = 329 \text{ cubic feet.}$$

Sch.—The following is Chezy's formula, with three different coefficients, varying from 69 "for small streams under 2000 cubic feet per minute," to 96 "for large rivers such as the Clyde or the Tay."

$v = 69 \, (r \sin \theta)^{\frac{1}{2}}$. For small streams.

$v = 93 \, (r \sin \theta)^{\frac{1}{2}}$. Eytelwein's coefficient.

$v = 96 \, (r \sin \theta)^{\frac{1}{2}}$. For large streams.

124. Bottom Velocity at which Scour Commences.—A river channel is said to have a fixed *regimen*, when it changes little in draft or form in a series of years. In some rivers, the deepest part of the channel changes its position perpetually, and is seldom found in the same place two successive years. The sinuousness of the river also changes by the erosion of the banks, so that in time the position of the river is completely altered. In other rivers, the change from year to year is very small, but probably the regimen is never perfectly fixed except where the rivers flow over a rocky bed. If a river had a constant discharge, it would gradually modify its bed till a permanent regimen was established. But as the volume discharged is constantly changing, and therefore the velocity, silt is deposited when the velocity decreases, and scour goes on when the velocity increases in the same place.

It has been found by experiment * that a stream moving with a velocity of 3 inches per second will carry along *fine clay* and *soft earth ;* moving 6 inches per second, will carry *loam ;* 1 foot per second, will carry *sand ;* 2 feet per second, *gravel ;* 3½ feet, *pebbles* an inch in diameter ; 4 feet, *broken stone, flint ;* 5 feet, *chalk, soft shale ;* 6 feet, *rock in beds ;* 10 feet, *hard rock.*

125. Transporting Power of Water.

—The specific gravity of rocks varies from 2.25 to 2.64; when immersed in water, therefore, they lose nearly half their weight. This fact greatly increases the transporting power of water. The pressure of a current of water against any surface varies as the square of the velocity and as the area of the surface † (Art. 97). But in similar figures, surfaces vary as the squares of the diameters; hence, the pressure of the current varies as the square of the velocity and as the square of the diameter, *i. e.*, the pressure of the current against a surface varies as the square of its velocity multiplied by the square of the diameter of the surface. Calling P the pressure which the current exerts against a rock, v its velocity, and d the diameter of the surface of the rock, we have

$$P \propto v^2 \times d^2. \qquad (1)$$

Now the *resistance* to be overcome, or the *weight* of the rock, varies as the cube of the diameter; *i. e.*, calling W the weight of the rock, we have

$$W \propto d^3. \qquad (2)$$

---

* Experiments by Dubuat. See Ency. Brit., Vol. XII., p. 503.

† Supposing that the area of the cross-section of the stream is at least large enough to cover the surface.

But when the current is just able to move the rock, we have

$$P \propto W. \quad (3)$$

Therefore, from (1), (2), and (3),

$$d^3 \propto v^2 \times d^2;$$

$$\therefore d \propto v^2,$$

which in (1) gives $P \propto v^2 \times v^4,$

or $\qquad P \propto v^6. \quad (4)$

That is, *the transporting power of a current varies as the sixth power of the velocity.*

This may also be shown geometrically as follows: Let a represent a cubic inch of stone, which a current of given velocity will just move, and let b be a cube of stone 64 times as large. Now if the velocity of the current be doubled, the force against each square inch of b will be four times as great as that against a; but the surface of b opposed to the current is sixteen times as great as that of a, and the pressure would be increased sixteen times from this cause; therefore the whole pressure against b from these two causes would be $4 \times 16 = 64$ times as great as against a.

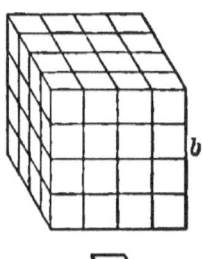

Fig. 65

But the weight also of b is 64 times as great as that of a; therefore the current would be just able to move it.

We have seen (Art. 124) that a current $3\frac{1}{2}$ feet per second, or about two miles an hour, will move pebbles an inch in diameter, or about three ounces in weight. It follows from the above law that a current of ten miles an hour will bear fragments of 1½ tons, and a torrent of 20 miles an hour will carry fragments of 100 tons in weight.*

---

* Le Conte's Geology, p. 18.

126. Back Water.

When a dam is built across a stream so as to raise the water and form a pond, the surface of the water in the pond will not be horizontal. Let AB represent

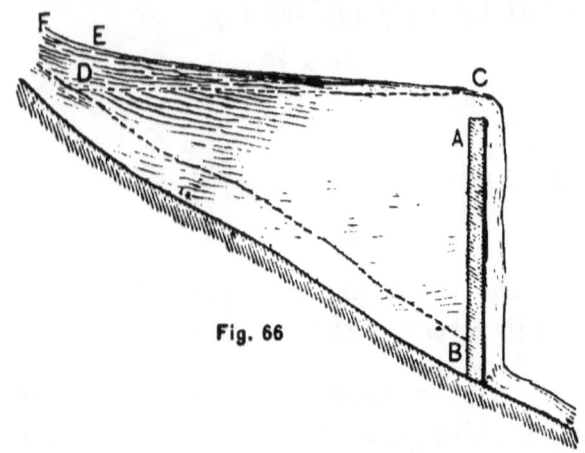

Fig. 66

a dam, and C the surface of the water directly over the dam. If the horizontal line CD be drawn from the surface at C to the point D, where it intersects the natural surface of the stream, the surface of the water in the pond will be everywhere above this line, except at C, its height increasing as the distance from the dam increases, and this elevation may extend for quite a distance up the stream above the point D.

The elevation CDFE above the horizontal CD is called *back water*. As the stream approaches the horizontal surface DC, its velocity is diminished, because the slope on which the velocity depends is very small, and as the velocity is diminished the water is heaped up above DC, even extending up the stream, until the slope is sufficient for the water to flow off. When this slope is established, the stream FEC flows smoothly along its liquid channel.

Fig. 66 shows a longitudinal section of the river Weser, in Germany, where a dam was built. The mean depth of the stream was about 2.5 feet, the surface was raised 7.5 feet, the slope of the stream was quite uniform for a distance of ten miles. At the point D, three miles from the dam, it was found by measurement that the surface E was elevated over 15 inches above D. At a distance of 4 miles above the dam, the surface was elevated by the dam 9 inches.*

127. River Bends.—When rivers flow in narrow valleys, where the banks do not readily yield to the action of the current, the effect of any variation of velocity is only temporarily to deepen the bed. In wide valleys and alluvial plains, where the soil of the banks is more easily worn by the current than the bottom, any increase in the volume of the water will widen the bed; and if one bank yields more than the other, *windings* or *bends* will be formed, and these windings which are thus formed tend to increase in curvature by the scouring away of material from the outer bank and the deposition of detritus along the inner bank. The windings sometimes increase till a loop is formed, with only a narrow strip of land between the two encroaching branches of the river. Finally, a "cut-off" may occur, a waterway being opened through the strip of land, and the loop left separated from the stream, forming a lagoon of marsh shaped like a horse-shoe.

It is usually supposed that the water, tending to go forwards in a straight line, rushes against the outer bank and scours it, at the same time creating deposits at the inner bank. This view is considered by many engineers as very incomplete. Prof. James Thomson has given an explanation of the action at a bend, which he has completely confirmed by experiment.† He thinks that the scouring at the outer side and the deposit at the inner side of the bend are due to the centrifugal force, in virtue of which the water passing round the bend presses outwards, and the free surface in a radial cross-section has a slope from the inner side upwards to the outer side.

---

* D'Aubuisson's Hydraulics, Art. 166.
† Proc. Inst. of Mech. Engineers, 1879, p. 456.

EXAMPLES.

1. A thin plane area moves edgeways through the water, in which it is completely immersed. Find the resistance per square foot at a speed of 20 miles per hour.

Ans. 3.442 lbs.

2. The length of a pipe is 400 feet, the head of water is 6 feet, and the diameter of the pipe is 6 inches, the entrance to it being cylindrical. Find (1) the head due to friction, (2) the velocity of discharge, and (3) the quantity discharged per second.

Take $f = 0.63$, $g = 32$, and use (3) of Art. 104 for v.

Ans. (1) 5.646 ft.; (2) 3.88 ft.; (3) 0.7619 cu. ft.

3. When the pipe is 800 feet long, the head of water 12 feet, and the diameter 6 inches, find (1) the friction head, (2) the velocity of discharge, and (3) the quantity discharged per second.

Ans. (1) 11.635 ft.; (2) 3.939 ft.; (3) 0.7734 cu. ft.

4. When the pipe is 1600 feet long, the head 24 feet, and the diameter 6 inches, find the same quantities as in the last two examples.

Ans. (1) 23.63 ft.; (2) 3.969 ft.; (3) 0.7793 cu. ft.

5. When the pipe is 3200 feet long, the head 48 feet, and the diameter 6 inches, find the same quantities.

Ans. (1) 47.627 ft.; (2) 3.984 ft.; (3) 0.7823 cu. ft.*

6. When the pipe is 800 feet long, the head 12 feet, and the diameter 5 inches, find the same three quantities as before.

Ans. (1) 11.694 ft.; (2) 3.605 ft.; (3) 0.4915 cu. ft.

---

* An inspection of Exs. 2, 3, 4, and 5, shows that if a 6-inch pipe be laid with a uniform slope of 6 feet in 400 feet, nearly all the head is consumed by friction, so that only a very small fraction of the entire head remains to generate the final velocity and to overcome the resistance at the entrance to the pipe, *i. e.*, in each case there is only about 0.35 of a foot of head left; one-third of this is expended in overcoming the resistance at the entrance to the pipe, and the other two-thirds in producing velocity.

7. When the length is 1600 feet, the head 24 feet, and the diameter 5 inches, find the same quantities.
Ans. (1) 23.69 ft.; (2) 3.628 ft.; (3) 0.4947 cu. ft.

8. When the length is 800 feet, the head 5 feet, and the diameter 6 inches, find the same quantities.
Ans. (1) 4.848; (2) 2.542; (3) 0.499.

9. When the length is 800 feet, the head 16 feet, and the diameter 6 inches, find the same quantities.
Ans. (1) 15.514; (2) 4.548; (3) 0.893.

10. Two pipes of the same length are 3 inches and 4 inches in diameter, respectively. Compare the losses of head by friction, (1) when the velocity is the same, and (2) when they deliver the same quantities of water.
Ans. (1) 1.33; (2) 4.21.

11. Water is to be raised to a height of 20 feet by a pipe 30 feet long and 6 inches in diameter. What is the greatest admissible velocity of the water, if not more than 10 per cent. additional power is to be required in consequence of the friction of the pipe? *Ans.* $8\frac{1}{2}$ feet per second.

12. Two reservoirs are connected by a pipe 6 inches in diameter and $\frac{3}{4}$ of a mile long. For the first quarter mile the pipe slopes at 1 in 50, for the second at 1 in 100, while in the third it is level. The head of water over the inlet is 20 feet, and that over the outlet 9 feet. Neglecting all loss except that due to surface friction, find (1) the velocity per second, and (2) the discharge in gallons per minute, assuming $f = 0.0348$. *Ans.* (1) 3.43 ft.; (2) 253 gallons.

13. A tank of 250 gallons is 50 feet above the street. It is connected with the street main, the head of which is 52 feet, by a pipe 100 feet long. (1) Find the diameter of the pipe that the tank may be filled in 20 minutes; (2) what must the head in the main be to fill the tank in 5 minutes with the pipe? *Ans.* (1) 1.6 inches; (2) 82 feet.

EXAMPLES. 239

14. What is the discharge per second through a pipe 48 feet long and 2 inches in diameter, under a head of 5 feet?

Sug.—Assume $f = .02$, and obtain from (8) of Art. 106, $v = 6.6$ feet, and therefore (Sch. of Art. 105), $f = .0211$, which in (8) of Art. 106 gives $v = 6.52$ feet. ∴ etc.

Ans. 245.8 cubic inches.

15. What must be the diameter of a pipe 100 feet long, which is to discharge one-half of one cubic foot of water per second under a head of 5 feet? *Ans.* 3.82 inches.

See remark in Art. 107.

16. If the diameter of one portion of the compound pipe (Fig. 55) is twice that of the other, and if the velocity of the water in the larger is 10 feet, find (1) the coefficient of resistance, and (2) the loss of head at the sudden enlargement, the water flowing from the small pipe into the large one. *Ans.* (1) 4; (2) 13.95 feet.

17. A pipe 2 inches in diameter is suddenly enlarged to 3 inches. If it discharge 100 gallons per minute, the water flowing from the small pipe into the large one, find (1) the coefficient of resistance, and (2) the loss of head at the sudden enlargement. *Ans.* (1) 1.59; (2) $8\frac{1}{2}$ inches.

18. In the last example, if the water moves in the reverse direction, find the loss of head caused by the sudden contraction, assuming the coefficient of contraction to be 0.66.

Ans. $7\frac{1}{2}$ inches.

19. A pipe contains a diaphragm with an orifice in it, the area of which is one-fifth the sectional area of the pipe. Find the coefficient of resistance of the diaphragm, assuming the contraction on passing through the orifice the same as that at efflux from a vessel through a small orifice in a thin plate. *Ans.* 46.

20. A horizontal pipe 30 feet long is suddenly enlarged from 2 inches to 3 inches, and then suddenly returns to its original diameter; the length of each section is 10 feet. If

240   EXAMPLES.

it discharge 100 gallons per minute into the atmosphere, find the total loss of head, assuming the coefficient of friction $f = .03$.   *Ans.* 10 ft. $2\frac{1}{2}$ ins.

21. Find the loss of head in inches due to a bend in a pipe 2 inches in diameter, the radius of curvature being 6 inches, and the velocity of the water being 12 feet per second.   *Ans.* 0.2 of an inch.

22. If the pipes in Ex. 2, Art. 106, which are to discharge 25 cubic feet of water per minute, contain two elbows, each of 90°, find the total loss of head.

Here we have, $h = (1.505 + 8.748 + 2 \times 0.984)\dfrac{v^2}{2g} =$ etc.

Ans. 1.76 feet.

23. If the pipe in Ex. 14 contains 5 bends, the radius of curvature of each being 2 inches, find (1) the velocity of the water issuing from the pipe, and (2) the quantity discharged per second.

Here β is found [from (2) of Art. 111] to be 0.294. $\therefore$ etc.

Ans. (1) 5.964 feet; (2) 224.81 cubic inches.

24. What quantity of water will be delivered by a canal 5800 feet long, when the fall is 3 feet, its depth 5 feet, its lower breadth 4 feet, and its upper breadth 12 feet?

Here $\dfrac{p}{a} = 0.42015$. $\therefore$ etc.

Ans. 129.48 cubic feet.

25. Find the quantity of water that is carried by a stream 40 feet wide, whose mean depth is $4\frac{1}{2}$ feet, and whose wetted perimeter is 46 feet, when it falls 1 inch in 75 feet.

Here v approximately, from (4) of Art. 121, $= 6.1$ feet. $\therefore f = 0.00765$, and v more correctly $= 6.05$ feet. $\therefore$ etc.

Ans. 1089 cubic feet.

26. A main 3100 feet long is to discharge water from a reservoir having a head of 75 feet. It is proposed to put in a 6-inch pipe for 800 feet, beginning at the reservoir, then a

5-inch pipe for 800 feet more, and a 4-inch pipe for the remaining 1500 feet. The coefficient of friction f being 0.024, find (1) the diameter of a uniform pipe 3100 feet long having the same friction, and (2) the velocity of discharge.

Use (7) of Art. 104 for v.

Ans. (1) 4.427 inches ; (2) 4.874 feet.

27. If in the last example the first pipe of the main is 2000 feet long and 6 inches in diameter, the second 800 feet long and 5 inches in diameter, and the third 300 feet long and 4 inches in diameter, the head being 75 feet, find (1) the diameter of the equivalent main, and (2) the velocity of discharge. *Ans.* (1) 5.2 inches ; (2) 5.289 feet.

28. If in Ex. 26 the pipe is 6 inches in diameter for the whole length of 3100 feet, the head being 75 feet, find the velocity of discharge. *Ans.* 5.675 feet.

29. Into a branch main 2000 feet long, 6 inches in diameter, water enters with a velocity of 15.27 feet a second; 1 cubic foot of water is delivered into service-pipes for every 1000 feet of length, $f = .0303$, what is the loss of head in the 2000 feet ? *Ans.* 211.53 feet.

CHAPTER III.

MOTION OF ELASTIC FLUIDS.

128. Work of the Expansion of Air.—If air expands without doing any work its temperature remains constant.* It follows from this that as air changes its state, the internal work done is proportional to the change of temperature. When, in expanding, air does work against an external resistance, either heat must be supplied or the temperature falls.

Suppose a given mass of air to be confined in a cylinder having a piston of one square foot area. Let v_1 be the initial volume and p_1 the initial pressure of the air, and suppose the piston to move so as to expand the air to any other volume v with pressure p. Then if heat is supplied to the air during the expansion so that the temperature remains constant, we have (Art. 48),

$$pv = p_1 v_1. \qquad (1)$$

Now if we represent the pressures by the ordinates, and the corresponding volumes by the abscissas of a curve AB referred to the axes OX, OY, the curve represents the relative changes of volume and pressure. Then OM $=$ v_1 and $MP_1 = p_1$ is a point P_1 corresponding to a volume v_1 and pressure p_1. Similarly (v, p) is any other point P of the curve corresponding to a volume v and pressure p; and since each member of (1) is constant, the curve is a *rectangular hyperbola*.

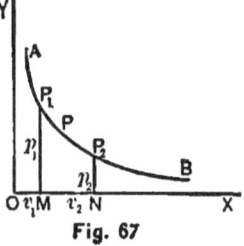

Fig. 67

---

* This result was first demonstrated experimentally by Joule.

The work of expansion between the pressures p_1 and p_2 is represented by the area of the space MP_1P_2N (Anal. Mechs., Art. 222). To find an algebraic expression for this work, let p and v be the corresponding pressure and volume at any intermediate point P in the expansion. Then the work done on the piston during the expansion from v to $v+dv$ is $p\,dv$, and the whole work done during the expansion from v_1 to v_2, represented by the area MP_1P_2N,

$$= \int_{v_1}^{v_2} p\,dv = p_1 v_1 \int_{v_1}^{v_2} \frac{dv}{v} \text{ [from (1)]}$$

$$= p_1 v_1 \log \frac{v_2}{v_1} = p_1 v_1 \log^* \frac{p_1}{p_2}, \qquad (2)$$

which is the work of expanding a given mass of air from a higher pressure p_1 to a lower pressure p_2.

Cor.—In order to compress a given mass of air whose volume is v_2 and whose pressure is p_2, into a volume v_1 of the pressure p_1, the work to be done

$$= p_2 v_2 \log \frac{p_1}{p_2}, \qquad (3)$$

which is the work of compressing a given mass of air from a lower pressure p_2 to a higher pressure p_1.

Sch.—The expressions in (2) and (3) for the work done during the expansion and compression of air, are correct only when the temperature of the air remains constant while the change of volume or density is taking place; but the temperature of the air remains constant only when the change of volume takes place so slowly that the heat in the confined air has sufficient time to communicate any excess to the walls of the vessel and to the exterior air. If the change of density occurs so quickly that it is accompanied by a change of temperature, when the air is expanded the

---

* Hyp. log.

temperature is lowered, and when the air is compressed the temperature is increased. Under these circumstances the pressure cannot change according to Art. 48, and other formulæ have to be produced. (See Weisbach's Mechs., p. 936; also Ency. Brit., Vol. XII., p. 480.)

129. Velocity of Efflux of Air According to Mariotte's Law.—Let the air be discharged from an orifice with the velocity v feet per second; let $w =$ the weight of a cubic foot of air, and $v_1 =$ the volume of air discharged per second; then the work performed by the volume of air v_1 in passing from the pressure p_1 to the pressure p_2, is, by (2) of Art. 128,

$$p_1 v_1 \log \frac{p_1}{p_2},$$

and this must be equal to the work stored in the air during the efflux, which is

$$\frac{v_1 w v^2}{2g}.$$

Therefore, we have

$$\frac{v_1 w v^2}{2g} = p_1 v_1 \log \frac{p_1}{p_2},$$

$$\therefore v = \sqrt{2g \frac{p_1}{w} \log \frac{p_1}{p_2}}. \tag{1}$$

A cubic foot of air, at the temperature 0° of the centigrade thermometer, and at a pressure corresponding to the height of 29.92 inches of the barometer, weighs about 0.08076 lbs.* Therefore for any temperature t, we have for the weight of a cubic foot of air, from (2) of Art. 54, since for the same volume and temperature the weight varies as the density,

$$w = \frac{0.08076}{1 + at}. \tag{2}$$

---

* Determined by Regnault. See Weisbach's Mechs., p. 795.

If the pressure differs from the mean pressure, or if the height of the barometer is not 29.92 inches, but b, (2) becomes

$$w = \frac{0.08076}{1 + at} \frac{b}{29.92}$$

$$= \frac{0.002699\, b}{1 + at}. \tag{3}$$

If we express the elastic force or pressure of the air by the pressure p upon each square inch, then we have

$$\frac{b}{29.92} = \frac{p}{14.7}, \tag{3a}$$

which in (3) gives

$$w = \frac{0.005494\, p}{1 + at}, \tag{4}$$

$$\therefore \frac{p}{w} = \frac{(1 + at)\, 144}{0.005494}, \tag{5}$$

where p is the pressure on each square foot.

Substituting this value in (1), we have

$$v = 161.9 \sqrt{2g\,(1 + at)\, \log \frac{p_1}{p_2}}. \tag{6}$$

If b is the height of the barometer and h that of the manometer (Art. 46), we have

$$\frac{p_1}{p_2} = \frac{b + h}{b}, \tag{6a}$$

which in (6) gives

$$v = 1299 \sqrt{(1 + at)\, \log\left(\frac{b + h}{b}\right)}, \tag{7}$$

where v is the velocity in feet, b the height of the barometer in the exterior air, h the height of the manometer for the air inside the vessel, t the temperature of the latter in degrees centigrade, and $a = 0.003665$, the coefficient of expansion of air (Art. 53).

Cor. 1.—If the pressures p_1 and p_2 are nearly equal to each other, we can put

$$\log\left(\frac{b+h}{b}\right) = \log\left(1 + \frac{h}{b}\right) = \frac{h}{b} - \tfrac{1}{2}\frac{h^2}{b^2},$$

which in (7) gives

$$v = 1299\sqrt{(1+at)\left(1-\frac{h}{2b}\right)\frac{h}{b}}. \qquad (8)$$

When $\frac{h}{b}$ is very small (8) becomes

$$v = 1299\sqrt{(1+at)\frac{h}{b}}. \qquad (9)$$

Cor. 2.—Taking $g = 32$, we have from (1)

$$v = 8\sqrt{\frac{p_1}{w}\log\frac{p_1}{p_2}}. \qquad (10)$$

When the pressures differ but little from each other, we may obtain approximate formulæ as follows: From (10), we have

$$v = 8\sqrt{\frac{-p_1}{w}\log\frac{p_2}{p_1}}. \qquad (11)$$

By development we have

$$\log\frac{p_2}{p_1} = \log\left(1 + \frac{p_2-p_1}{p_1}\right)$$

$$= \frac{p_2-p_1}{p_1} - \tfrac{1}{2}\left(\frac{p_2-p_1}{p_1}\right) + \text{etc.}$$

Neglecting all powers of $\frac{p_2-p_1}{p_1}$ above the first, and substituting in (11), we have

$$v = 8\sqrt{\frac{p_1-p_2}{w}}. \qquad (12)$$

Neglecting all powers of $\dfrac{p_2-p_1}{p_1}$ above the second, we have

$$v = 8\sqrt{\dfrac{p_1-p_2}{w}\left(1+\dfrac{p_1-p_2}{2p_1}\right)}. \qquad (13)$$

If h be the height of a homogeneous fluid, of the same density as the air, which is necessary to produce the pressure p_1-p_2, then $p_1-p_2 = wh$, which in (12) gives

$$v = 8\sqrt{h}. \qquad (14)$$

It will be observed that (8), (9), (12), (13), (14) are true only when the pressures p_1 and p_2 are nearly equal to each other.

130. Efflux of Moving Air.—*To find the velocity of efflux when the pressure of the air is given in the pipe through which it flows.*

The formulæ for efflux found in Art. 129, are based upon the supposition that the pressure p, or the height h, of the manometer is measured at a place where the air is at rest, or moving very slowly. If the pressure be measured at a point where the air is in motion, in determining the velocity of efflux, we must take into account the kinetic energy of the moving air.

Let p_1 be the pressure of the air in the pipe A, as indicated by the manometer M, and v_1 the velocity of the air passing the orifice of the manometer; p_2 the pressure of the air at efflux, and v_2 its velocity; a_1 the area of the section of the pipe A, and a_2 the area of the orifice O; w the

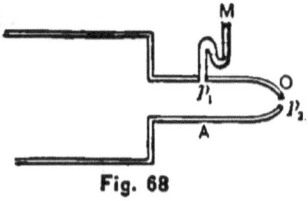

Fig. 68

weight of a cubic foot of the air, and Q the volume discharged per second.

Then the work stored in the air while passing from the pipe A to the orifice O

$$= \frac{Qw}{2g}(v_2^2 - v_1^2);$$

and this must equal the work done by the expansion of the air from p_1 to p_2. Therefore, from (2) of Art. 128, we have

$$\frac{Qw}{2g}(v_2^2 - v_1^2) = p_1 Q \log \frac{p_1}{p_2}. \tag{1}$$

The volume of air passing through the pipe per second is $a_1 v_1$, and that which passes the orifice is $a_2 v_2$; hence we have (Art. 48)

$$a_1 v_1 \times p_1 = a_2 v_2 \times p_2; \quad \therefore \quad v_1 = \frac{a_2 p_2}{a_1 p_1} v_2,$$

which in (1) and reducing, gives

$$v_2 = \sqrt{\frac{2gp_1 \log \frac{p_1}{p_2}}{w\left[1 - \left(\frac{a_2 p_2}{a_1 p_1}\right)^2\right]}}, \tag{2}$$

which is the velocity of efflux.

Cor.—Substituting for the numerator its value as given in (8) of Art. 129, we have

$$v_2 = 1299 \sqrt{\frac{(1 + at)\left(1 - \frac{h}{2b}\right)\frac{h}{b}}{1 - \left(\frac{a_2 p_2}{a_1 p_1}\right)^2}}; \tag{3}$$

or approximately, when p_1 is not much greater than p_2,

$$v_2 = 1299 \sqrt{\frac{(1 + at)\frac{h}{b}}{1 - \left(\frac{a_2}{a_1}\right)^2}}. \tag{4}$$

131. Coefficient of Efflux.

When air issues from an orifice, the section of the current undergoes a contraction similar to that observed in the efflux of water (Art. 91). If the orifice of efflux is in a thin plate, the stream of air has a smaller cross-section than the orifice, and the practical discharge is less than the theoretical.

Denoting the *coefficient of contraction* by c, we have, as in the case of water (Art. 92), $c =$ the ratio of the cross-section of the stream of air to that of the orifice.

Denoting the *coefficient of velocity* by ϕ, we have, as in Art. 93, $\phi = \dfrac{v_1}{v}$, where v_1 is the actual and v the theoretical velocity of discharge.

Denoting the *coefficient of efflux* by μ, we have, as in Art. 94, $\mu = \dfrac{Q_2}{Q_1} = c\phi$, where Q_2 is the actual and Q_1 the theoretical discharge.

The older experiments upon the efflux of air through orifices vary considerably from each other. According to the experiments of Koch,* $\mu = 0.58$ when the air issues from an orifice in a thin plate; $\mu = 0.74$ when the air issues from a pipe about six times as long as it is wide; and $\mu = 0.85$ when the air issues from the conical nozzle of a bellows about five times as long as it is wide and having a lateral convergence of $6°$. D'Aubuisson, Poncelet, and Pecqueur found values somewhat different.†

Weisbach has found the following values of the coefficient of efflux μ: ‡

Conoidal mouth-pieces of the form of the contracted vein, with effective pressures of 0.23 to 1.1 atmosphere, $\mu =$ 0.97 to 0.99

---

* Tate's Mech. Phil., p. 329.   † Weisbach's Mechs., p. 945.
‡ Ency. Brit., Vol. XII., p. 481.

Circular, sharp-edged orifices, ...   0.563 to 0.788
Short, cylindrical mouth-pieces, . . .   0.75
Conical pipes, whose angle of convergence
 is about 6°, ...   0.92
Conical converging mouth-pieces, well
 rounded off, ...   0.98

132. The Quantity Discharged.—Let k be the area of the orifice in square feet, and Q_2 the discharge per second in cubic feet at the pressure of the external air. Then we have, from (1) of Art. 129,

$$Q_2 = \mu k \sqrt{2g \frac{p_1}{w} \log \frac{p_1}{p_2}}$$

$$= 1299 \mu k \sqrt{(1 + \alpha t) \log \left(\frac{b + h}{b}\right)} \quad (1)$$

[from (7) of Art. 129];

or, from (8) and (9) of Art. 129, we have

$$Q_2 = 1299 \mu k \sqrt{(1 + \alpha t)\left(1 - \frac{h}{2b}\right)\frac{h}{b}} \quad (2)$$

$$= 1299 \mu k \sqrt{(1 + \alpha t)\frac{h}{b}}. \quad (3)$$

When Q_2 is reduced to the pressure of the air inside the vessel, we have, from Art. 48,

$$Q_1 p_1 = Q_2 p_2;$$

$$\therefore Q_1 = 1299 \mu k \frac{b}{b+h} \sqrt{(1 + \alpha t) \log \left(\frac{b+h}{b}\right)}. \quad (4)$$

From (4) of Art. 130, we have

$$Q = 1299 \mu k \sqrt{\frac{(1 + \alpha t)\frac{h}{b}}{1 - \left(\frac{a_2}{a_1}\right)^2}}. \quad (5)$$

133. Coefficient of Friction of Air.

When air flows through a long pipe, it has, like water, *a resistance of friction* to overcome, due to the surface of the pipe; and this resistance, which is found to consume by far the greater part of the work expended, can be measured by the height of a column of air, which is determined by the expression,

$$h = f \frac{l}{d} \frac{v^2}{2g}, \qquad (1)$$

in which, as in the case of water (Art. 103), l denotes the length, d the diameter of the pipe, v the velocity of the air, and f the coefficient of resistance of friction, to be determined by experiment. The work expended in friction generates heat, the most of which must be developed in the air and given back to it. Some heat may be transmitted through the sides of the pipe to surrounding materials, but, in all the experiments that have thus far been made, the amount so conducted away appears to be very small; and if no heat is transmitted, the air in the pipe must remain sensibly at the same temperature during expansion; that is, the heat generated by friction exactly neutralizes the cooling due to the work done.

A discussion by Prof. Unwin [*] of the experiments by Messrs. Culley and Sabine on the rate of transmission of light carriers through pneumatic tubes, in which there is a steady flow of air not sensibly affected by any resistances other than the surface friction, furnished the value $f = 0.028$. The pipes were of lead, slightly moist, 2¼ inches (0.187 ft.) in diameter, and in lengths of 2000 to nearly 6000 feet.

Girard's experiments upon the motion of air in pipes gave a mean coefficient of resistance, $f = 0.0256$; those of D'Aubuisson gave as a mean, $f = 0.0238$; while those of Buff gave the mean value of $f = 0.0375$.

According to the experiments of Weisbach, it is only when velocities are about 80 feet that the coefficient of resistance can be put $= 0.024$, and it diminishes as the velocity of the air in the pipe increases.

---

[*] See Ency. Brit., Vol. XII., p. 491.

He found that the coefficient of friction, when the velocity was given in feet, could be expressed approximately by the following formula,

$$f = \frac{0.217}{\sqrt{v}}. \tag{2}$$

The resistance caused by *elbows* and *bends* is to be treated in the same way as in the case of water (Arts. 110, 111).

134. Motion of Air in Long Pipes.—By the aid of the coefficient of friction of a pipe, we can calculate the velocity of efflux and the discharge for a given length and diameter of the pipe.

Let $v =$ the velocity of discharge $= \dfrac{Q}{k}$.

$v_1 =$ the velocity of the air in the pipe.
$d =$ the diameter of the orifice, whose area therefore is $k = \frac{1}{4}\pi d^2$.
$d_1 =$ the diameter of the pipe.
$\beta_0 =$ the coefficient of resistance at the entrance to the pipe.
$f =$ the coefficient of resistance due to the friction of the pipe.
$\beta_1 =$ the coefficient of resistance at the orifice.
$p_1 =$ the pressure of the air when it is discharged.
$w =$ the weight of a cubic foot of air.
$h =$ the height of the manometer in the reservoir.
$b =$ the height of the barometer.
$l =$ the length of the pipe.

Then the height due to the resistance at the entrance to the pipe,

$$= \beta_0 \frac{v_1^2}{2g} = \beta_0 \frac{d^4}{d_1^4} \frac{v^2}{2g}.$$

The height due to the resistance of friction in the pipe

$$= f \frac{l}{d_1} \frac{v_1^2}{2g} = f \frac{l}{d_1} \frac{d^4}{d_1^4} \frac{v^2}{2g}.$$

The height due to the resistance at the orifice

$$= \beta_1 \frac{v^2}{2g}.$$

The height due to the velocity

$$= \frac{v^2}{2g}.$$

Therefore, the total height

$$= \left[\left(\beta_0 + f\frac{l}{d_1}\right)\frac{d^4}{d_1^4} + 1 + \beta_1\right]\frac{1}{2g}\left(\frac{Q}{k}\right)^2. \quad (1)$$

Also, the total height, from (1) and (6a) of Art. 129,

$$= \frac{p_1}{w} \log\left(1 + \frac{h}{b}\right) = \frac{p_1}{w}\frac{h}{b}, \text{ approximately.} \quad (2)$$

Therefore, equating (1) and (2), and solving for Q, we have

$$Q = k \sqrt{\frac{2g \frac{p_1}{w}\frac{h}{b}}{\left(\beta_0 + f\frac{l}{d_1}\right)\left(\frac{d}{d_1}\right)^4 + 1 + \beta_1}} ;$$

which, from (9) of Art. 129,

$$= 1299\frac{\pi d^2}{4}\sqrt{\frac{(1 + \alpha t)\frac{h}{b}}{\left(\beta_0 + f\frac{l}{d_1}\right)\left(\frac{d}{d_1}\right)^4 + 1 + \beta_1}}, \quad (3)$$

where $\beta_0 = \frac{1}{\mu_0^2} - 1$, and $\beta_1 = \frac{1}{\mu_1^2} - 1$ (Art. 96).

Sch.—In Paris, Berlin, London, and other cities, it has been found cheaper to transmit messages in pneumatic tubes than to telegraph by electricity. The tubes are laid under ground, with easy curves; the messages are made into a roll and placed in a light felt carrier, the resistance of which in the tubes in London is only ¾ oz. A current of air, forced into the tube or drawn through it, propels the carrier. In most systems the current of air is steady and continuous, and the carriers are introduced or removed without materially altering the flow of air.

135. The Law of the Expansion of Steam.—When steam is produced in a close vessel, as in the boiler of a steam engine, the density of the steam increases with the temperature; but so long as the temperature remains the same, the quantity of steam that can be raised from the water is limited, and the steam is generated at its maximum density and pressure for the temperature, whatever this may be; if the temperature falls, a portion of the steam resumes the liquid form, and the density of the steam is diminished. When the steam is in its condition of maximum density, it is said to be saturated, being incapable of vaporizing or absorbing more water into its substance, or increasing its pressure, so long as the temperature remains the same. Also, on the contrary, steam will not be generated with less than the maximum quantity of water which it is capable of appropriating from the liquid out of which it ascends. Any change in either one of the three elements of pressure, density, or temperature of steam is necessarily accompanied by a change of the other two. The same density is invariably accompanied by the same pressure and temperature.

If the volume of steam over water be increased, while the temperature remains constant, then, as long as there is liquid in excess to supply fresh vapor to occupy the increased space, the density will not be diminished, but will remain constant with the pressure. If the source of heat be removed, when all the liquid is evaporated, the pressure and density will diminish, when the volume is increased, as in permanent gases; and if the volume be again diminished, the pressure and density will increase, until they return to the maximum due to the temperature; and the effect of any further diminution of volume, or attempt to further increase the density at the same temperature, is simply accompanied by the precipitation of a portion of the vapor to the liquid state, the density remaining the same.

On the contrary, if the application of heat be continued when all the liquid is evaporated, the state of saturation ceases, and the temperature and pressure are increased, while the density remains the same; the steam is said to be superheated, or surcharged with heat, and it becomes more perfectly gaseous. While in this condition, if it were to be replaced in contact with water of the original temperature, it would evaporate a part of the water, transferring to it the surcharge of heat, and would resume its normal state of saturation.

If the space for steam over the water remain unaltered, then, if the temperature is raised by the addition of heat, the density of the vapor is increased by fresh vaporization, and the elastic force is consequently increased in a much more rapid ratio than it would be in a permanent gas by the same change of temperature. Conversely, if the temperature be lowered, a part of the vapor is condensed, the density is diminished, and the elastic force reduced more rapidly than in a permanent gas. The density of saturated steam is about $\frac{5}{8}$ of that of atmospheric air, when they are both under the same pressure and at the same temperature.

It has been determined experimentally that whatever may be the pressure at which steam is formed, the quantity of fuel necessary to evaporate a given volume of water is always the same; also the relation between the temperature and pressure of saturated steam has been determined experimentally, and from this tables have been formed giving the relation between the pressure and volume of steam raised from a cubic foot of water.* Since the volume is always a function of the pressure, we may write

$$V = f(P). \tag{1}$$

136. Work of Expansion of Steam.—Let V be the volume of steam from a cubic foot of water at the pressure P, where P is the pressure on a square foot, and U_Q the work performed by Q cubic feet of water, in the form of steam, between the pressures P and P_1; let ABCD be a vertical section of the space in which the steam expands, ABPQ the volume V of the steam at the pressure P, ABP_1Q_1 the volume V_1 of the steam at pressure P_1, A the area of the section at PQ in feet, and dv the distance between the two consecutive sections PQ and MN.

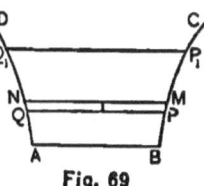

Fig. 69

Then, for the element of work performed by one cubic foot of water in the form of steam at the pressure P, we have

$$dU_1 = AP\, dv = P dV,$$

since $A dv = dV$. Integrating between the limits P and P_1, we have

$$U_1 = \int_{P_1}^{P} P\, dV = \int_{P_1}^{P} P df(P) \tag{1}$$

[from (1) of Art. 135].

Therefore, for the work done by Q cubic feet of water in expanding from P to P_1, we have

$$U_Q = Q \int_{P_1}^{P} P df(P), \tag{2}$$

which can be integrated when the function $f(P)$ is known.

SCH.—Since from (2) the quantity of work done is entirely independent of the form of the vessel ABCD, it

---

* See Ency. Brit., Art. *Steam*.

follows that the work of steam between any given pressures P and P_1 is always the same, whatever may be the nature of the space through which the steam expands, and that it increases with the pressure P at which the steam is generated; since therefore the quantity of fuel necessary to evaporate a given volume of water is always independent of the pressure at which the steam is formed (Art. 135), it follows that it is most economical to employ steam of as high a temperature as possible.

137. Work of Steam at Efflux.—Let w be the weight of Q cubic feet of water evaporated per second, V the volume of steam from a cubic foot of water at the pressure P, which is the pressure of the steam at the point of efflux, k the section of the orifice in feet from which the steam is discharged, and v the velocity per second. Then calling U_Q the work stored in the steam at efflux, we have

$$U_Q = \frac{w}{2g} v^2. \tag{1}$$

Since $V = f(P)$, and $QV = kv$, we have

$$kv = Qf(P) = \frac{w}{62.5} f(P);$$

$$\therefore v = \frac{w}{62.5 k} f(P), \tag{2}$$

which in (1) gives

$$U_Q = \frac{w^3}{7812.5 k^2 g} [f(P)]^2. \tag{3}$$

Hence, *the work of steam discharging itself from an orifice varies as the cube of the water evaporated.*

Cor. 1.—The work stored in the steam at efflux is due to the work of expansion between the pressures P and P_1;

therefore, from (2) of Art. 136 and (1) of the present Art., we have

$$\frac{w}{2g}v^2 = Q\int_{P_1}^{P} Pdf(P) = \frac{w}{62.5}\int_{P_1}^{P} Pdf(P); \quad (4)$$

$$\therefore v = \mu\left[\frac{2g}{62.5}\int_{P_1}^{P} Pdf(P)\right]^{\frac{1}{2}}, \quad (5)$$

which gives the velocity of efflux, μ being the coefficient of efflux (Art. 131).

Cor. 2.—From (1) of Art. 130, we have

$$U_Q = \frac{62.5\,Q}{2g}(v_2^2 - v_1^2). \quad (6)$$

Let V_1 and V_2 be the volumes of steam per second from a cubic foot of water at P_1 and P_2 pressures respectively; then (Art. 130),

$$QV_1 = a_1 v_1, \qquad QV_2 = a_2 v_2,$$

$$\therefore v_1 = Q\frac{V_1}{a_1}, \qquad v_2 = Q\frac{V_2}{a_2},$$

which in (6) gives

$$U_Q = \frac{62.5\,Q^3}{2g}\left[\left(\frac{V_2}{a_2}\right)^2 - \left(\frac{V_1}{a_1}\right)^2\right], \quad (7)$$

and we see again, as in (3), that *the work of steam varies as the cube of the water evaporated.*

Solving (7) for $Q\dfrac{V_2}{a_2} = v^2$, we have

$$v^2 = \left[\frac{2gU_Q}{62.5\,Q} + Q^2\left(\frac{V_1}{a_1}\right)^2\right]^{\frac{1}{2}}, \quad (8)$$

which gives the theoretical velocity of efflux.

WORK OF STEAM IN THE EXPANSIVE ENGINE. 259

138. Work of Steam in the Expansive Engine.— Let K be the area of the piston in square feet, h_1 the length of the stroke, including the clearance,* h the point of the cylinder at which the steam is cut off, Q the number of cubic feet of water evaporated per minute, P and P_1 the pressures of the steam at the beginning and end of the stroke respectively, N the number of strokes performed by the piston per minute, V the volume of steam from a cubic foot of water at P pressure, and $V = f(P)$, as before. Then, for the volume of steam discharged per minute at P pressure, we have

$$Qf(P) = NKh. \qquad (1)$$

Similarly, $\quad Qf(P_1) = NKh_1,$

$$\therefore \frac{f(P)}{f(P_1)} = \frac{h}{h_1},$$

and $\qquad f(P_1) = \frac{h_1}{h} f(P). \qquad (2)$

The work performed upon the piston before the steam is cut off is $NKP(h-c)$; adding this to (2) of Art. 136, we have

$$U_Q = Q \int_{P_1}^{P} P \, df(P) + NKP(h-c)$$

$$= Q \left[\int_{P_1}^{P} P \, df(P) + \frac{h-c}{h} P f(P) \right] \qquad (3)$$

[from (1)],

which is the total work performed by the steam per minute.

Cor.—Let L be the useful load in lbs. upon each square foot of the piston, F the friction in lbs. per square foot of the piston, arising from the motion of the unloaded piston, f the coefficient of friction arising from the useful load, and

---

* The *clearance* is the space in the cylinder lying beneath the piston, at the lowest point of its stroke.

p the pressure of the steam in the condenser. Then the total resistance upon the piston is $K[F + L(1 + f) + p]$, and therefore the work expended per minute in overcoming this resistance

$$= NK[F + L(1 + f) + p](h_1 - c). \qquad (4)$$

When the mean motion of the piston of the engine is uniform, the work of the resistance will be equal to the work of the steam; therefore, by equating (3) and (4), and reducing by (1), we have

$$\frac{h}{f(P)}\int_{P_1}^{P} P\, df(P) + (h - c) P$$
$$= [F + L(1 + f) + p](h_1 - c), \qquad (5)$$

from which the value of the useful load L is readily determined.

EXAMPLES.

1. If a blowing machine changes per second 10 cubic feet of air, at a pressure of 28 inches, into a blast at a pressure of 30 inches, find the work to be done in each second.

Here p_2, from (3a) of Art. 129, $= 0.4913b$; $\therefore$ etc.

Ans. 1366.7 foot-lbs.

2. If under the piston of a steam engine, whose area is 201 square inches, there is a quantity of steam 15 inches high and at a pressure of 3 atmospheres, and if this steam in expanding moves the piston forward 25 inches, find the work of the expansion per second. *Ans.* 10866 foot-lbs.

3. The air in a reservoir is at a temperature of 120° C., and at a pressure corresponding to a height of the manometer of 5 inches, while the barometer marks 29.2 inches. Find (1) the theoretical velocity of efflux, and (2) the theoretical discharge through an orifice 1½ inches in diameter.

Use (9) of Art. 129.

Ans. (1) 645.12 feet; (2) 7.917 cubic feet.

EXAMPLES. 261

4. The height of a manometer, which is placed upon a pipe $3\frac{1}{4}$ inches in diameter through which the air is passing, is $2\frac{1}{2}$ inches, while the air is discharged through an orifice 2 inches in diameter at the end of the pipe. Find (1) the theoretical velocity of efflux, and (2) the theoretical discharge, if the barometer in the external air stands at $27\frac{1}{2}$ inches, and the air in the pipe is at a temperature of 10° C.

Use (5) of Art. 132.

Ans. (1) 421.8 feet ; (2) 9.2 cubic feet.

5. If in the last example the height of the manometer is 3 inches, the diameter of the pipe is 4 inches, and the orifice at the end of the pipe is 1 inch in diameter, find (1) the velocity, and (2) the discharge when the barometer stands at 29 inches and the temperature of the air in the pipe is 20° C. *Ans.* (1) 447.06 feet; (2) 2.438 cubic feet.

6. If the sum of the areas of two conical tuyeres of a blowing machine is 3 square inches, the temperature in the reservoir is 15°, the height of the manometer in the regulator is 3 inches, and the height of the barometer in the exterior air is 20 inches, find the discharge.

See (3) of Art. 132 ; take $\mu = .92$, and $a = .004$.*

Ans. 8.242 cubic feet.

7. The height of a quicksilver manometer, which is placed upon a regulator at the head of an air pipe 320 feet long and 4 inches in diameter is 3.1 inches, the height of the barometer in the free air is 29 inches, the diameter of the orifice in the conically convergent end of the pipe is 2 inches, and the temperature of the compressed air in the regulator is 20° C. Find the quantity of air that is delivered through this pipe.

Sug. $a = 0.004$, $\mu_0 = .75$, $\mu_1 = .92$; ∴ etc.

Ans. 5.735 cubic feet.

---

* On account of the ordinary humidity of the atmosphere, it is advisable in practice to take $a = 0.004$.

8. If the height of the manometer in the last example is 4 inches, the pipe 500 feet long and 6 inches in diameter, the height of the barometer 30 inches, the diameter of the orifice in the conically convergent end of the pipe 2 inches, and the temperature of the compressed air in the regulator 30° C., find the quantity discharged.

Ans. 9.051 cubic feet.

9. If the height of the manometer in Ex. 7 is 2.5 inches, the pipe 600 feet long and 5 inches in diameter, the height of the barometer 29.5 inches, the diameter of the orifice in the conically convergent end of·the pipe one inch, and the temperature of the compressed air in the regulator is 10° C., find the quantity discharged. *Ans.* 1.883 cubic feet.

CHAPTER IV.

HYDROSTATIC AND HYDRAULIC MACHINES.

139. Definitions.—There are several simple machines whose action depends on the properties of air and water; a brief description of some of these machines will now be given, sufficient to exhibit the principles involved in their construction and use.

Hitherto the energy exerted by means of a head of water has been wholly employed in overcoming frictional resistances, and in generating the velocity with which the water is delivered at some given point. In the cases which we have now to consider, only a fraction of the head is required for these purposes; the remainder, therefore, becomes a source of energy at the point of delivery by means of which useful work may be done.

Hydraulic energy may exist in three forms, according as it is due to *motion*, *elevation*, or *pressure*. In the first two cases the energy is inherent in the water itself, being a consequence of its motion or position, as in the case of any other heavy body. In the third it is due to the action of gravity or some other force, sometimes on the water itself, but oftener on other bodies; the water then only transmits the energy, and is not directly the source of it.*

140. The Hydrostatic Bellows.—This machine presents an illustration of the principle of the transmission of fluid pressure (Art. 8). It consists of a cylinder CDEF (Fig. 70), with its sides made of leather or other flexible material, and a pipe ABF leading into it. If water is

---
* Cotterill's App. Mechs., p. 482.

poured into the pipe till the vessel and pipe are filled, a very small pressure applied at A will raise a very great weight upon DE, the weight lifted being greater as DE is greater.

Let k be the area of a horizontal section of the pipe, K that of a section of the cylinder, or that of DE, and p the pressure applied at A. Then, from (1) of Art. 9, we have

$$\frac{p}{W} = \frac{k}{K}. \qquad (1)$$

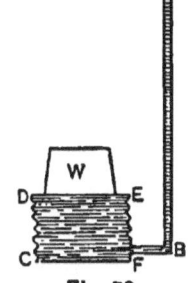

Fig. 70

SCH.—Suppose the pipe AB to be extended vertically upwards, and the pressure at A to be produced by means of a column of water above it, formed by pouring in water to a considerable height, and suppose the pipe to be very small, so that the pressure upon the section A may be very small; then, as this pressure is transmitted to every portion of the surface DE that is equal to the section A, the upward force produced on DE can be as large as we please. To increase the upward force, we must enlarge the surface DE or increase the height of the column of water in the pipe, and the only limitation to the increase of the force will be the want of sufficient strength in the pipe and cylinder to resist the increased pressure. By making the pipe AB of very small bore, and the height DC of the cylinder very small, the quantity of water can be made as small as we please. That is, *any quantity of fluid, however small, may be made to support any weight, however great.* This is known as *the hydrostatic paradox.*

141. The Siphon.—The action of a *siphon* is an important practical illustration of atmospheric pressure. It is simply a bent tube of unequal branches, open at both ends, and is used to convey a liquid from a higher to a lower level, over an intermediate point higher than either.

THE SIPHON.

Let A and B be two vessels containing water, B being on the lower level, and ACB a bent tube. Suppose this tube to be filled with water from the vessel A, and to have its extremities immersed in the water in the two vessels. The water will then flow from the vessel A to B, as long as the level B is below A, and the end of the shorter branch of the siphon is below the surface of the water in the vessel A.

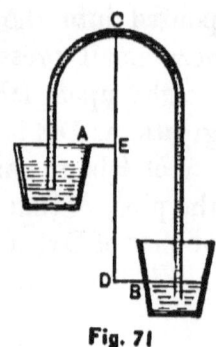

Fig. 71

The atmospheric pressures upon the surfaces A and B tend to force the water up the two branches of the tube. When the siphon is filled with water, each of these pressures is counteracted in part by the pressure of the water in the branch of the siphon that is immersed in the water upon which the pressure is exerted. The atmospheric pressures are very nearly the same for a difference of level of several feet, owing to the slight density of air. The pressures of the suspended columns of water, however, will for the same difference of level differ considerably, in consequence of the greater density of water. The atmospheric pressure opposed to the weight of the longer column will therefore be more resisted than that opposed to the weight of the shorter, thereby leaving an excess of pressure at the end of the shorter branch, which will produce the motion. Thus, draw the vertical line DEC, let h denote the height of the water barometer, k the area of a section of the tube, and w the weight of a unit of volume; then the water at the point C is urged from left to right by a force

$$= wkh - wk \times EC;$$

and it is urged from right to left by a force

$$= wkh - wk \times DC.$$

Subtracting the second from the first, we have

$$wk\,(DC - EC) = wk \times DE,$$

for the resultant force which urges the water at C from left to right, and hence there will be a continuous flow of water from the upper to the lower vessel.

It will be observed that the direction of the flow is wholly due to the fact that the level of the water in B is below that of the water in A. It is not necessary therefore that the longer branch should be immersed in the water; so long as the end B of the tube is below the water surface in A, the water will continue to flow through the tube ACB, until either the surface in A has fallen below the end of the tube, or, if the siphon be long enough, until the surface in A has descended so far that its depth below C is greater than h.

Sch.—The siphon is often used to drain ponds, marshes, and canals, and when used for this purpose it is made of leather, or stout canvas, like the common hose.

142. The Diving Bell.—This is a large bell-shaped vessel made of iron, open at the bottom, and containing seats for several persons. Its weight is greater than that of the water it would contain, and when lowered by a chain into the water, the air which it contains becomes more and more compressed as it sinks, in consequence of the increasing pressure to which it is subject. As the volume of air diminishes the water rises in the bell; but the air will prevent the water from rising high in the bell, and the persons seated within are thus enabled to descend to considerable depths and to carry on their operations in safety. When the surface of the water within the bell is at a depth of 33 feet below the outer surface the bell will be half filled with water. The bell is supplied with fresh air from above by a flexible tube connected with an air pump, and may be entirely emptied of water by the air forced in by the pump. There are also contrivances for the expulsion of the air when it becomes impure.

The force tending to lift the bell is the weight of the water displaced by the bell and the enclosed air. Hence the tension on the suspending chain, being equal to the weight of the bell diminished by the weight of water displaced by the bell and the air within, will increase as the bell descends, in virtue of the diminution of air space due to the increased pressure, unless fresh air is forced in from above.

Let ABCD be the bell, let $EF = a$, the depth of its top below the surface of the water, $FK = b$, the height of the cylinder, $FH = x$, the length occupied by air, π and π' the pressures of the atmospheric air and of the compressed air within the bell, and h the height of the water barometer. Then we have (Art. 48)

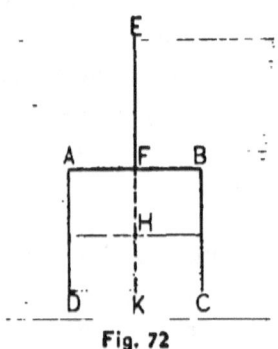

Fig. 72

$$\pi' = \pi \frac{b}{x} = \pi + g\rho(a + x). \qquad (1)$$

But, $\pi = g\rho h,$

which in (1) gives

$$x^2 + (a + h)x = hb,$$

$$\therefore x = \frac{-(a + h) + \sqrt{(a + h)^2 + 4bh}}{2}. \qquad (2)$$

the positive value only being the one which belongs to the problem.

Cor.—If A be the area of the top of the bell, and its thickness be neglected, the volume of displaced water is Ax, and the tension of the chain

$$= \text{weight of bell} - g\rho Ax. \qquad (3)$$

Sch.—The principle of the diving bell is applied in diving dresses. The diver is clothed in a water-tight dress fitted with a helmet, and is supplied with air by means of a pump. There is an escape valve by which the circulation of fresh air is maintained. The diver may be weighted up to 200 lbs., but on closing the escape valve, he can rise at once to the surface in virtue of the buoyancy due to the increased displacement of water by the enclosed air.

143. The Common Pump (Suction Pump).—Any machine* used for raising water from one level to a higher, in which the agency of atmospheric pressure is employed, is called a *pump*. Pumps are either *suction, forcing*, or *lifting* pumps.

The pump most commonly in use is a *suction* pump, of which Fig. 73 is a vertical section. AB and BC are two cylinders connected together having a common axis; the former is called the barrel of the pump and the latter the suction pipe; M is a piston accurately fitting the barrel, and movable up and down through the space AB by means of a vertical rod EV, connected with a handle or lever EF, which turns on a fulcrum O; in the piston is a valve V which opens upwards, and at the top of the suction pipe BC is another valve V′, which likewise opens upwards. S is a spout a little above A, and C is the surface of the water in which the lower part of the pump is immersed.

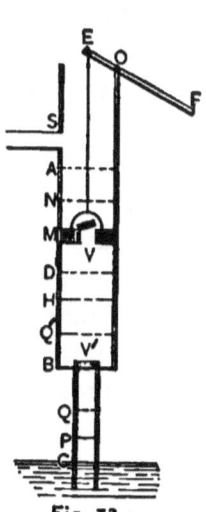

Fig. 73

To explain the action of the suction pump, suppose the piston M to be at B, the pump filled with ordinary atmos-

---

* Machines for raising water have been known from very early ages, and the invention of the common pump is generally ascribed to Ctesibius, teacher of the celebrated Hero of Alexandria ; but the true theory of its action was not understood till the time of Galileo and Torricelli. (See Deschanel's Nat. Phil., p. 215.)

pheric air, and the valves V and V' closed by their own weight; the water will stand at the same level C both within and without the suction pipe. Now raise the piston, the air in BC will tend by its elastic force to occupy the space which the piston leaves void; it will therefore open the valve V', and will pass from the pipe to the barrel, its elasticity diminishing in proportion as it fills a larger space. It will, therefore, exert less pressure on the water at C than the atmosphere does at C outside the pump; hence the atmospheric pressure on the surface of the water outside will force water up the pipe BC, until the pressure at C is equal to the atmospheric pressure. As the piston rises the water will rise in BC, the pressure of the air above M keeping the valve V closed. When the piston descends, the valve V' closes, and the air in MB, becoming compressed as the piston descends, will at length have its elastic force greater than that of the exterior air above the piston, and will open the valve V, and will escape through it.

This process being repeated a few times, the water at length ascends through the valve V' into the barrel, and at the next descent of the piston, will be forced through the valve V and be then lifted to the spout S, through which it will flow. While this water is being lifted, the atmospheric pressure on the surface of the water outside the pipe forces more water into the pump, so that, on the next descent, the piston gets more water to lift; and thus the process continues, the suction pipe and barrel remaining full, so that a cylinder of water equal to that through which the piston is raised will be poured out at each upward motion, provided the spout S is large enough.

SCH. 1.—The height BC must be less than the height of the water barometer, or else the water will never rise to the valve V'. Although the height of the water barometer is about 33 feet, yet in consequence of unavoidable imperfections in construction, the height of the valve V' above the surface of the water in the well should be considerably less than 33 feet; otherwise the quantity of water lifted by the piston at each stroke will be small.

Sch. 2.—It is not essential to the construction that there should be two cylinders; a single cylinder, with a valve somewhere below the lowest point of the piston-range will be sufficient, provided the lowest point of the range be less than 33 feet above the surface in the reservoir.

It is not necessary to the working of a pump that the suction pipe should be straight; it may be of any shape, and may enter the reservoir at any horizontal distance below the barrel of the pump.

144. Tension of the Piston Rod.—(1) If the water in BC (Fig. 73) has risen to P when the piston is at M, let π' be the pressure of the air in MP; then we have $\pi' =$ pressure of water at P = pressure of water at C $- g\rho$PC; hence

$$\pi' = \pi - g\rho\text{PC}. \qquad (1)$$

But the tension on the rod is the difference between the atmospheric pressure above the piston and the pressure of the air in MP; hence calling A the area of the piston and T the tension of the rod, we have from (1)

$$\text{T} = (\pi - \pi')\text{A} = g\rho\text{PC}\cdot\text{A}. \qquad (2)$$

If one inch be taken as the unit of length, and h be the height in inches of the water barometer, we have $g\rho h = 15$ lbs., nearly, which in (2) gives

$$\text{T} = 15\frac{\text{PC}\cdot\text{A}}{h}. \qquad (3)$$

(2) *When the pump is in full action.*—Let AH be the range of the piston, and let CD = h, then at each stroke, the volume DH of water is lifted, and therefore the tension of the rod when the piston is ascending will be $g\rho\text{A}.(h + \text{HD})$ until the water begins to flow through the spout.

Therefore in the suction pump the tension of the rod is equal to the weight of the **column of water** whose base is the area of the piston and whose height is the height of the water in the pump above the level of the well.

If A be on a level with the spout, all the water lifted will be discharged, and as the piston descends, the tension of the rod will be $g\rho A h$.

145. Height Through which the Water Rises in One Piston Stroke.—Let P and Q (Fig. 73) be the surfaces of the water at the beginning and end of an upward stroke of the piston from B to A, and let h as usual be the height of the water barometer. The air which occupied the space BP at the beginning of the stroke occupies at the end of it the space AQ; and the pressures are respectively

$$g\rho(h - \text{PC}), \quad g\rho(h - \text{QC}).$$

Hence (Art. 48)

$$h - \text{PC} : h - \text{QC} :: \text{vol. AQ} : \text{vol. BP}.$$

If R and r are the radii of the cylinders AB and BC, we have

$$\text{vol. AQ} = \pi R^2 \text{AB} + \pi r^2 (\text{BC} - \text{QC}),$$

$$\text{vol. BP} = \pi r^2 (\text{BC} - \text{PC}),$$

$$\therefore \frac{h - \text{PC}}{h - \text{QC}} = \frac{R^2 \text{AB} + r^2 (\text{BC} - \text{QC})}{r^2 (\text{BC} - \text{PC})},$$

which determines QC for any given value of PC.

Cor.—If the stroke of the piston be less than AB, as for instance AH, then HC must be less than h. Also, a limit exists with regard to H, which may be shown as follows:

If P be the surface of the water when the piston M is at A, then, as the piston descends, the valve V′ will close, but the valve V will not be opened until the pressure of the air in MB is greater than the atmospheric pressure. When M is at A the pressure of the air $= g\rho\,(h - PC)$, and unless the valve V is opened before M arrives at H, the pressure of the air in HB will

$$= g\rho\,(h - PC)\,\frac{AB}{HB},$$

which must be greater than $g\rho h$, if the valve is to open, and therefore $h \cdot AH$ must be greater than $AB \cdot PC$. Hence, to insure the opening of the valve while the surface is below B, we must have

$$h \cdot AH > AB \cdot BC, \qquad (1)$$

or
$$\frac{AH}{AB} > \frac{BC}{h};$$

i. e., the ratio of AH to HB must be at least as great as the ratio of BC to h. This condition, although necessary in every case, may not be sufficient.

For, suppose that the surface of the water is at Q' when the piston M is at A, in which case the pressure of the air in $AQ' = g\rho\,(h - Q'C)$.

When the piston descends to H, the pressure in HQ'

$$= g\rho\,(h - Q'C)\,\frac{AQ'}{HQ'},$$

which must be greater than $g\rho h$, if the valve is to open, and therefore
$$h \cdot AH > AQ' \cdot Q'C.$$

But the greatest value of $AQ' \cdot Q'C$ is $\tfrac{1}{4}\overline{AC}^2$; therefore we must have

$$h \cdot AH > \tfrac{1}{4}AC^2. \qquad (2)$$

Since $\tfrac{1}{4}\overline{AC}^2 > AB \cdot BC$, unless B is the middle point of AC, it follows that the condition in (2) includes the condition in (1), which is therefore in general insufficient. (See Besant's Hydrostatics, p. 97.)

146. The Lifting Pump.—When water has to be raised to a height exceeding about 30 feet, the suction pump will not work (Art. 143, Sch. 1), and the lifting pump is commonly used. By means of this instrument, water can be *lifted* to any height. It consists of two cylinders, in the upper of which a piston M is movable, the piston-rod working through an airtight collar. A pipe DF is carried from the barrel to any required height; at D there is a valve which opens into the pipe. The suction pipe BC is closed by a valve V', as in the suction pump, and the piston M usually * has a valve V.

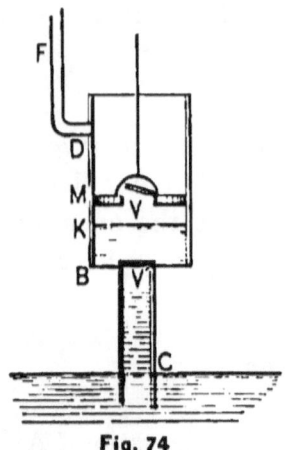

Fig. 74

The action of this pump is precisely the same as that of the suction pump in raising water from the well into the barrel. Suppose the piston at its highest point, and the surface of the water in the barrel at K ; then, as the piston is depressed, its valve V will open, and the water will flow through it till the piston reaches its lowest point. When the piston ascends, lifting the water, the valve D opens, and water ascends in the pipe DF. On the descent of the piston, the valve D closes, and every successive stroke increases the quantity of water in the pipe, until at last it is filled, after which every elevation of the piston will deliver a volume of water equal to that of a cylinder whose base is the area of the piston and whose height is equal to its stroke. The only limit to the height to which water can be lifted is that which depends on the strength of the instrument and the power by which the piston is raised.

COR.—If $CK = h$, the piston lifts the volume BK at

---

* Sometimes the piston has no valve in it, but is replaced by a solid cylinder, called a plunger, which is operated by a handle as before.

each stroke, and if $A =$ the area of the piston, the tension on the piston-rod $= g\rho A \cdot BK$, until the water is lifted to the valve D, since the air is expelled before the machine is in full action. After this, the power applied to the piston-rod must be increased until the pressure of the water opens the valve D, i. e., until the pressure $= g\rho\,(h + DF)$, where F is the surface of the water in the tube. The water will then be forced up the tube, the tension of the rod increasing as the surface F ascends.

147. The Forcing Pump.—This pump is a further modification of the simple suction pump; it has no valve in its piston, which is perfectly solid, and works water-tight in the barrel, ranging over the space AE. At the top of the suction pipe BC is a valve, and at the entrance to the pipe DF is a second valve D.

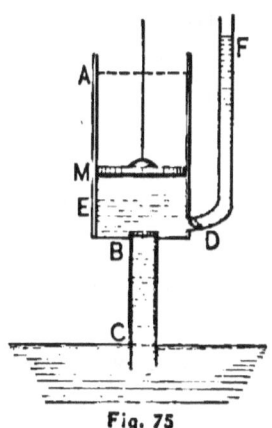

Fig. 75

When this pump is first set in action, water is raised from the well as in the common pump, by means of the valve B and piston M, the air at each descent of the piston being driven through the valve D into the pipe DF. When the water has risen through B, the piston, descending, forces it through D; and when the piston ascends, the valve D closes, and more water enters through B. The next descent of the piston forces more water through D, and so on until the pipe is filled, as in the lifting pump.

The stream which flows from the top of the pipe will be intermittent, as it is only on the descent of the piston that water is forced into the pipe; but a continuous stream can be obtained by means of a strong air vessel N (Fig. 76), which consists of a strong brass or copper vessel, at the bottom of which is a valve V. Through the top of the air

THE FORCING PUMP.

vessel is a discharge pipe KF, which passes air-tight nearly to the bottom. When water is forced into the air vessel through the valve V by the descent of the piston, it rises above the lower end of this pipe. The mass of air which the vessel contains is compressed into a smaller volume; its elastic force, pressing on the surface of the water at K, with a varying but continuous pressure, forces it up the pipe; and if the size of the vessel be suitable to that of the pump, and to the rate of working it, the compressed air will continue to expand, forcing water up the pipe during the ascent of the piston, and will not have lost its force before a new compression is applied to it, carrying with it a new supply of water, and thus a continuous, although varying, flow will be maintained. A few strokes of the piston will generally be sufficient to raise water in the pipe KF, to any height consistent with the strength of the instrument and the power at command.

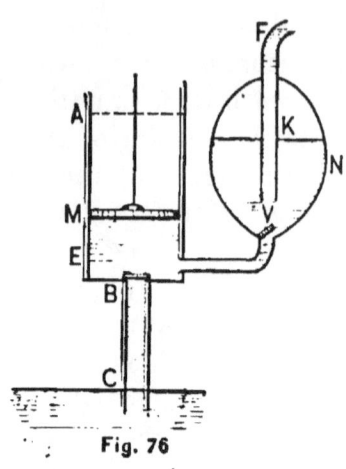

Fig. 76

COR.—Let h = the height of the water barometer; during the ascent of the piston the valve B is open and V is closed; the pressure upon the upper surface of the piston = gph; the pressure upon the lower surface = $gp(h - MC)$, the water surface in the pump being at M; therefore, calling A the area of the piston, the tension of the rod when the piston is ascending = $gp.A \cdot MC$.

That is, *the tension of the rod is equal to the weight of a column of water whose base is the area of the piston, and whose height is the height of the water in the barrel above the level of the well.*

148. The Fire Engine.—This is only a modification of the forcing pump with an air vessel, as just described.

Two cylinders M and M' are connected with the air vessel V by means of the valves D and D', and the pistons are worked by means of a lever GEG', the ends of which are raised and depressed alternately, so that one piston is ascending while the other is descending.

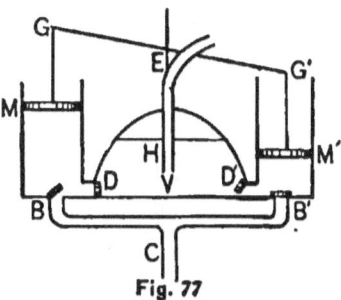

Fig. 77

Water is thus continually being forced out of the air vessel through the vertical pipe EH, which has a flexible tube of leather attached to it, by means of which the stream can be thrown in any direction.

149. Bramah's Press.[*]—This press is a practical application of the principle of the equal transmission of fluid pressures (Art. 8). In the vertical section of this instrument (Fig. 78), A and C are two solid pistons or cylinders fitting in air-tight collars, and working in the strong hollow cylinders L and K, which are connected by a pipe BD. At D is a valve opening upwards, and at B is a valve opening inwards, a pipe from D communicating with a reservoir of water. M is a movable platform, supporting the substance to be pressed, and N is the top of a strong frame. HOF is the lever working the cylinder C, F being the fulcrum, and H the handle.

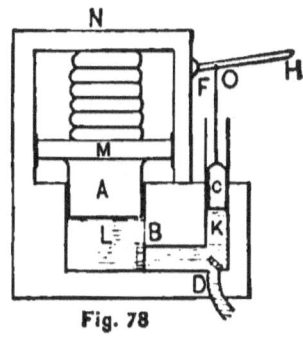

Fig. 78

Action of the Press.—Let C be raised; the atmospheric

---

[*] The principle of this press was suggested by Stevinus. It remained unfruitful in practice until 1796, when Bramah, an English engineer, by an ingenious contrivance, overcame the only difficulty which prevented its practical application.

pressure forces water from the reservoir through the valve D into the hollow cylinder K, as in the common pump. The cylinder C being pressed down, the valve D closes, and the water is forced through the valve B into L, and, acting on the cylinder A, makes it ascend, thus producing pressure upon any substance included between M and N. A continued repetition of this process will produce any required compression of the substance.

Let R and r be the radii of the cylinders A and C, p the power applied at the handle H, and P the pressure of the water on A; then we have, for the downward force p' on C,

$$p' = p \frac{\text{HF}}{\text{OF}}. \qquad (1)$$

But (Art. 9) $P : p' = R^2 : r^2$;

$$\therefore \; P = p \frac{\text{HF}}{\text{OF}} \frac{R^2}{r^2}. \qquad (2)$$

By increasing the ratio of R to r, any amount of pressure may be produced. Presses of this kind were employed in lifting into its place the Britannia Bridge over the Menai Straits, and for launching the Great Eastern.

150. Hawksbee's Air-Pump.*—B and B' are two cylinders, in which pistons P and P', with valves V and V' opening upward, are worked by means of a toothed wheel, the one ascending as the other descends. At the lower extremity of the cylinders there are valves v and v' opening upwards, and communicating by means of the pipe AC with the receiver R, from which the air is to be exhausted.

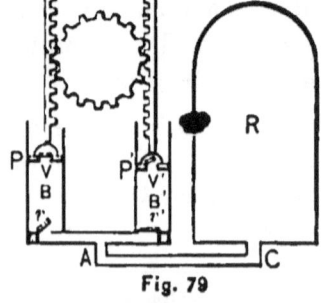

Fig. 79

---

* The air-pump was invented in 1650 by Otto von Guericke, Burgomaster of Magdeburg.

Suppose P at its lowest and P' at its highest position, and turn the wheel so that P ascends and P' descends. When P' descends, the valve v' closes and the air in B' flows through V', while the valve V is closed by the pressure of the external air, and air from R, by its elastic force, opens the valve v and fills the cylinder B. When P descends, the valve v closes, and the air in B being compressed flows through the valve V, while the valve V' closes, and air from the receiver flows through v' into B'. At every stroke of the piston, a portion of the air in the receiver is withdrawn; and after a considerable number of strokes a degree of rarefaction is attained, which is limited only by the weight of the valves which must be lifted by the pressure of the air beneath.

Let A denote the volume of the receiver, and B that of either cylinder; ρ the density of atmospheric air, and ρ_1, $\rho_2, \ldots \rho_n$ the densities in the receiver after $1, 2, \ldots n$ descents of the pistons. Then after the first stroke the air which occupied the space A will occupy the space $A + B$, and therefore we have

$$\rho_1 (A + B) = \rho A.$$

Similarly, $\qquad \rho_2 (A + B) = \rho_1 A ;$

$$\therefore \rho_2 (A + B)^2 = \rho A^2,$$

and after n strokes we have

$$\rho_n (A + B)^n = \rho A^n,$$

the volume of the connecting pipe AC being neglected.

Hence, calling π_n and π the pressures of the air in the receiver after n strokes and of the atmospheric air respectively, we have

$$\frac{\pi_n}{\pi} = \frac{\rho_n}{\rho} = \left(\frac{A}{A+B}\right)^n. \qquad (1)$$

Thus, suppose that A is four times B, and we were required to find the density of the air in the receiver at the end of the 15th stroke, we have from (1)

$$\rho_{15} = \rho \left(\tfrac{4}{5}\right)^{15} = 0.03552\,\rho.$$

If the air originally had an elastic force equal to the pressure of 30 in. of mercury, this would give the elastic force of the air remaining in the receiver as equal to a pressure of 1.056 in. of mercury. In this case, it is customary to say that the *vacuum pressure* is one of 1.056 in. of mercury.

SCH.—It is evident from (1) that ρ_n can never become zero as long as n is finite, and therefore, even if the machine were mechanically perfect, we could not by any number of strokes completely remove the air; for, after every stroke there would be a certain fraction left of that which occupied it before.

In working the instrument, the force required is that which will overcome the friction, together with the difference of the pressures on the under surfaces of the pistons, the pressures on their upper surfaces being the same.

151. Smeaton's Air-Pump.—This instrument consists of a cylinder AB in which a piston is worked by a rod passing through an air-tight collar at the top; a pipe BD passes from B to the glass receiver C, and three valves, opening upwards, are placed at B, A, and in the piston.

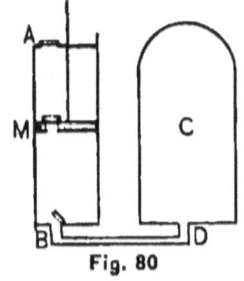

Fig. 80

Suppose the receiver and cylinder to be filled with atmospheric air, and the piston at B. Raising the piston, the valve A is opened by the compressed air in AM which flows out through it, while at the same time a portion of the air in C flows through the pipe DB to fill the partial vacuum formed in MB, so that when the piston arrives at A, the air which at first occupied C now fills both the receiver and the cylinder. When the piston descends, the valves A and B close, and the air in the cylinder below the piston is compressed until it opens the valve M, and passes above the piston. As the piston is raised a second time the valve A is opened by the compressed air in AM, which flows out through it as before; and thus at each stroke of the piston a portion of the air in the receiver is forced out through A.

Let A and B denote the volumes of the receiver and cylinder respectively, and ρ and ρ_n the densities of atmospheric air and of air in the receiver after n strokes. Then, as in Art. 150, we have

$$\rho_n (A + B)^n = \rho A^n,$$

from which it appears as in the previous article that, although the density of the air will become less and less at every stroke, yet it can never be reduced to nothing, however great n may be.

SCH.—An advantage of this instrument is that, the upper end of the cylinder being closed, when the piston descends the valve A is closed by the external pressure, and therefore the valve M is then opened easily by the air beneath. Also the labor of working the piston is diminished by the removal, during the greater part of the stroke, of the atmospheric pressure on M, which is exerted only during the latter part of the ascent of the piston, when the valve A is open.

152. The Hydraulic Ram.*—The hydraulic ram is a machine by which a fall of water from a small height produces a momentum which is made to force a portion of the water to a much greater height.

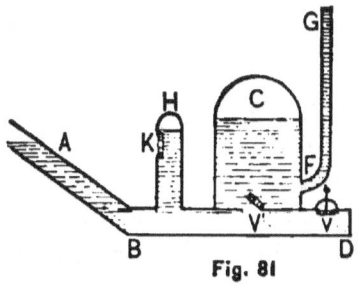

Fig. 81

---

* Invented by Montgolfier.

In the vertical section (Fig. 81), AB is the descending and FG the ascending column of water, which is supplied from a reservoir at A. V is a valve opening downwards, and V' is a valve opening upwards into the air-vessel C; H is a small auxiliary air-vessel with a valve K opening inwards.

The Action of the Machine.—As the valve V at first lies open by its own weight, a portion of the water, descending from A, flows through it; but the upward flow of the water towards the valve V increases the pressure tending to lift the valve, and at last, if the valve is not too heavy, lifts and closes it. The forward momentum of the column of water ABD being destroyed by the stoppage of the flow, the water exerts a pressure sufficient to open the valve V' and to flow through it into the air-vessel C, condensing the air within; the reaction of the condensed air forces water up the pipe FG. As the column of water ABD comes to rest, the pressure of the water diminishes, and the valves V and V' both fall. The fall of the former produces a rush of the water through the opening V, followed by an increased flow down the supply pipe AB, the result of which is again the closing of V, and a repetition of the process just described, the water ascending higher in FG, and finally flowing through G.

The action of the machine is assisted by the air-vessel H in two ways—first, by the reaction of the air in H, which is compressed by the descending water, and, secondly, by the valve K, which affords supplies of fresh air. When the water rises through V', the air in H suddenly expands, and its pressure becoming less than that of the outer air, the valve K opens, and a supply flows in, which compensates for the loss of the air absorbed by the water and taken up the column FG, or wasted through V. About a third of the water employed is wasted, but the machine once set in motion will continue in action for a long time, provided the supply in the reservoir be maintained. (See Besant's Hydrostatics, p. 112.)

153. Work of Water Wheels.—To utilize a head of water, consisting of an actual elevation above a datum level at which the water can be delivered and disposed of, a machine may be employed in which the direct action of the weight of the water, while falling through the given height is the principal moving force.

When a stream of water strikes the paddles of a wheel which has a certain velocity, the energy imparted to the wheel by the water, from (4) of Art. 98,

$$= [v^2 - (v - V)^2]\frac{W}{2g}, \qquad (1)$$

where V is the velocity of the periphery of the wheel, v the original velocity of the water, and W the weight of water acting on the wheel per second; but if the water descends with the paddle there is an additional amount of work done on the wheel due to the mean height h through which the water falls. Hence we have, for the whole work done on the wheel per second,

$$= [v^2 - (v - V)^2]\frac{W}{2g} + Wh. \qquad (2)$$

Now if the water leaves the paddles the work remaining in the water will be lost; hence, calling v_1 the velocity of the water after it has left the paddles, we have for the useful work U done on the wheel

$$U = [v^2 - (v - V)^2 - v_1^2]\frac{W}{2g} + Wh$$

$$= [2vV - V^2 - v_1^2]\frac{W}{2g} + Wh, \qquad (3)$$

which is the general expression for the work done by a water wheel when the water impinges upon the paddles perpendicularly.

154. Work of Overshot Wheels.

When a waterfall ranges between 10 and 70 feet, and the water supply is from 3 to 25 cubic feet per second, it is possible to construct a bucket wheel on which the water acts *chiefly* by its weight. If the variation of the head-water level does not exceed 2 feet, an overshot wheel may be used. The water is then projected over the summit of the wheel, and falls in a parabolic path into the bucket. If v be the velocity of delivery to the wheel,

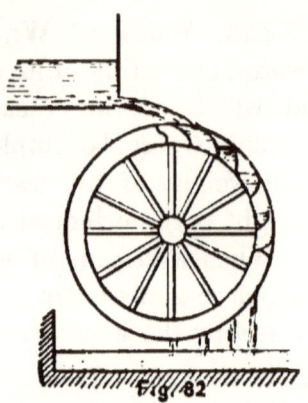

Fig. 82

the part $\dfrac{v^2}{2g}$ is converted into energy of motion before reaching the buckets and operates by impulse; hence in a wheel of this class the water does not operate *entirely* by weight.

The height h through which the water falls is the vertical height of the point at which the water meets the buckets above the point where it leaves them, which in this wheel is nearly equal to the diameter of the wheel; and as the velocity of the water on leaving the bucket is the same as the velocity of the bucket itself, we have $v_1 = V$; hence (3) of Art. 153 becomes

$$U = (v - V)V\frac{W}{g} + Wh. \qquad (1)$$

Calling m the efficiency* of these wheels, we have from (1)

$$U = m\left[\frac{1}{g}(v - V)V + h\right]W. \qquad (2)$$

Cor.—To find the relation of v and V so that the useful work U of the wheel may be a maximum, we must equate to zero the derivative of U with respect to V, which gives

---

* See Anal. Mechs., Art. 216.

$V = \tfrac{1}{2}v$, i. e., the wheel works to the best advantage when the velocity of its periphery is one-half that of the stream.

Sch.—If the velocity of the periphery of this wheel is too great, water is thrown out of the buckets before reaching the bottom of the fall. In practice, the circumferential velocity of water wheels of this kind is from $4\tfrac{1}{2}$ to 10 feet per second, about 6 feet being the usual velocity of good iron wheels not of very small size. The velocity of the water therefore is limited to about 12 feet per second, and the part of the fall operating by impulse is therefore about $2\tfrac{1}{4}$ feet. The rest of the fall operates by gravitation, but a certain fraction is wasted by spilling from the buckets, and emptying them before reaching the bottom of the fall. The great diameter of wheel required for very high falls is inconvenient, but there are examples of wheels 60 feet in diameter and more.

The efficiency of these wheels under favorable circumstances is 0.75, and is generally about 0.65.

155. Work of Breast Wheels.—When the variation of the head-water level exceeds 2 feet, a breast wheel is better than an overshot. In breast wheels the buckets are replaced by vanes which move in a channel of masonry partially surrounding the wheel. The water falls over the top of a sliding sluice in the upper part of the channel. The channel is thus filled with water, the weight of which rests on the vanes and furnishes the motive force on the wheel. There is a certain amount of leakage between the vanes and the sides of the channel, but this loss is not so great as that by spilling from the buckets of the overshot wheel.

Fig. 83

In this wheel, as in the case of the overshot wheel, $v_1 = V$, therefore (1) and (2) of Art. 154 also apply to breast wheels, h being the height of the point at which the water meets the vanes above the point where it leaves them. The efficiency is found by experience to be as much as 0.75.

Sch. 1.—Theoretically this wheel also works to the best advantage when the speed of its periphery is one-half that of the stream (Art. 154, Cor.). But Morin found, by experiments, that the efficiency of the wheel is not much affected by changes in its velocity. This is owing to the circumstance that the useful work is dependent principally upon the term Wh, and not upon the other term in the formula which alone is affected by the velocity of the wheel. Hence the great advantage of this wheel is, that it may be worked, without materially impairing its efficiency, with velocities varying from $\frac{1}{4}v$ to $\frac{3}{4}v$.

Sch. 2.—As the diameter of this wheel is greater than the fall, a breast wheel can be employed only for moderate falls.

Overshot and breast wheels work badly in back-water, and hence if the tail-water level varies, it is better to reduce the diameter of the wheel so that its greatest immersion in flood is not more than one foot.

156. Work of Undershot Wheels.

—The common undershot wheel consists of a wheel provided with vanes, against which the water impinges directly. In this case the water is allowed to attain a velocity due to a considera-

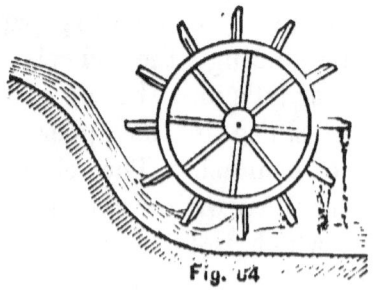

Fig. 64

ble part of the head immediately before entering the machine, so that its energy is nearly all converted into energy of motion; and as the water has no fall on the wheel, and its velocity on leaving the vanes is the same as the velocity of the vane itself, we have $h = 0$, $v_1 = V$; therefore (3) of Art. 153 becomes

$$U = (v - V) V \frac{W}{g}, \qquad (1)$$

or

$$U = m (v - V) V \frac{W}{g}, \qquad (2)$$

where m, as before, is the efficiency of the machine.

SCH.—The wheel works to the best advantage when the speed of the periphery is one-half that of the stream (Art. 154, Cor.), but the efficiency is low, never exceeding 0.5.

Wheels of this kind are cumbrous. In the early days of hydraulic machines, they were often used for the sake of simplicity. In mountain countries, where unlimited power is available, they are still found. The water is then conducted by an artificial channel to the wheel, which sometimes revolves in a horizontal plane. When of small diameter, their efficiency is still further diminished.*

157. Work of the Poncelet Water Wheel.—When the fall does not exceed 6 feet, the best water motor to adopt in many cases is the Poncelet undershot water wheel. In the common undershot water wheel, the paddles are flat, whereas in the Poncelet wheel they are curved, so that the direction of the curve at the lower edge, where the water first meets the paddle, is the same as the direction of the stream. By this arrangement, the water, which is allowed to flow to the wheel with a velocity nearly equal to the velocity due to the whole fall, glides up the curved floats without meeting with any sudden obstruction, comes to relative rest, then descends along the float, and acquires at the point of discharge from the float a backward velocity relative to the wheel nearly equal to the forward velocity of

---

* See Cotterill's App. Mechs.; also, Fairbairn's Millwork and Machinery.

WORK OF THE PONCELET WATER WHEEL.

the wheel. The water will therefore drop off the floats deprived of nearly all its kinetic energy. Nearly the whole of the work of the stream must therefore have been expended in driving the float; and the water will have been received without shock, and discharged without velocity.

Let v and V be the velocities of the stream and float respectively; then the initial velocity of the stream relative to the float is $v - V$, and the water will continue to run up the curved float until it comes to relative rest; it will then descend along the float, acquiring in its descent, under the influence of gravity, the same relative velocity which it had at the beginning of its ascent, but in a contrary direction. Therefore the absolute velocity of the water leaving the float is $V - (v - V) = 2V - v$.

Now the useful work U done on the wheel must equal the work stored in the water at first, diminished by the work stored in the water on leaving the wheel; hence

$$U = \frac{W}{2g} v^2 - \frac{W}{2g} (2V - v)^2$$

$$= \frac{2W}{g} (v - V) V. \tag{1}$$

Comparing this expression with (1) of Art. 156, we see that the work performed by the Poncelet wheel is double that of the common undershot wheel.

SCH.—This wheel works to the best advantage when the speed of the periphery is one-half that of the stream (Art. 154, Cor.). This conclusion also follows from the form of the floats, as above described; since if all the work is taken out of the water when it leaves the floats, its velocity must then be zero, and therefore $2V - v = 0$, or $V = \frac{1}{2}v$.*

The efficiency of a Poncelet wheel has been found in ex-

---

* The inventor, Poncelet, states that, in practice, the velocity of the water, in order to produce its maximum effect, ought to be about 2½ times that of the wheel, and that the efficiency of the wheel is about 0.7 (Tate's Mech. Phil., p. 313).

periments to reach 0.68. It is better to take it at 0.6 in estimating the power of the wheel, so as to allow some margin.

158. The Reaction Wheel; Barker's Mill.—

Fig. 85 shows a simple reaction wheel. ACB is a tube, capable of revolving about its axis, which is vertical, and having a horizontal tube DBE connected with it. Water is supplied at C, which descends through the vertical tube, and issues through the orifices D and E at the extremities of the horizontal tube, so placed that the direction of motion of the water is tangential to the circle described by the orifices. The efflux is in opposite directions from the two orifices; as the water flows through BD, the pressures on the sides balance each other except at D, where there is an uncompensated pressure on the side opposite the orifice; the effect of this pressure or reaction is to cause motion in a direction opposite to that of the jet. The same effect is produced by the water issuing at E, and a continued rotation of the machine is thus produced by the reaction of the jet in each arm.

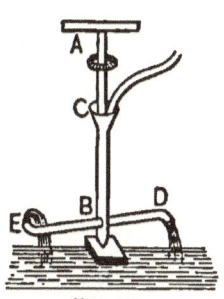

Fig. 85

Let h be the available fall, measured from the level of the water in the vertical pipe to the centres of the orifices, v the velocity of discharge through the jets, and V the velocity of the orifices in their circular path. When the machine is at rest, the water issues from the orifices with the velocity $\sqrt{2gh}$ (neglecting friction). But when the machine rotates, we have for the velocity of discharge through the orifices, from (1) of Art. 89,

$$v = \sqrt{V^2 + 2gh}. \qquad (1)$$

While the water passes through the orifices with the velocity v, the orifices themselves are moving in the opposite direction with the velocity V. The absolute velocity of the water is therefore

$$v - V = \sqrt{V^2 + 2gh} - V. \qquad (2)$$

Now the useful work done per second by each pound of water must equal the work due to the height h, diminished by the work remaining in the water after leaving the machine. Hence,

$$\text{useful work} = h - \frac{(v - V)^2}{2g}$$

$$= \frac{(\sqrt{V^2 + 2gh} - V)\,V}{g}, \text{ from (2), } (3)$$

$$= \frac{(v - V)\,V}{g}. \qquad (4)$$

The whole work expended by the water fall is h foot-pounds per second; consequently, to find the efficiency of the machine, we divide (3) by h (Anal. Mechs., Art. 216), and get

$$\text{efficiency} = \frac{(\sqrt{V^2 + 2gh} - V)\,V}{gh} \qquad (5)$$

$$= 1 - \frac{gh}{2V^2} + \text{etc.} \qquad (6)$$

(by the Binomial Theorem),

which increases towards the limit 1 as V increases towards infinity. Neglecting friction, therefore, the maximum efficiency is reached when the wheel has an infinitely great velocity of rotation. But this condition is impracticable to realize; and even at practicable but high velocities of rotation, the prejudicial resistances, arising from the friction of the water and the friction upon the axis, would considerably reduce the efficiency. Experiment seems to show that the best efficiency of these machines is reached when the velocity is that due to the head, so that $V^2 = 2gh$.

When $V^2 = 2gh$, we have, from (5), neglecting friction,

$$\text{efficiency} = \frac{(\sqrt{2}-1) V^2}{gh} = 0.828, \qquad (7)$$

about 17 per cent. of the energy of the fall being carried away by the water discharged. The actual efficiency realized of these machines appears to be about 60 per cent., so that about 22 per cent. of the whole head is spent in overcoming frictional resistances, in addition to the energy carried away by the water.

SCH.—The reaction wheel in its crudest form is a very old machine known as "*Barker's Mill.*" It has been employed to some extent in practice as an hydraulic motor, the water being admitted below and the arms curved. In this case the water is transmitted by a pipe which descends beneath the wheel and then turns vertically upwards. The vertical axle is hollow, and fits on to the extremity of the supply pipe with a stuffing box. In this construction the upward pressure of the water may be made equal to the weight of the wheel, so that the pressure upon the axis may be nothing. These modifications do not in any way affect the principle of the machine, but the frictional resistances may probably be diminished.

159. The Centrifugal Pump.—When large quantities of water are to be raised on a low lift, no pump is so suitable as a centrifugal pump. In this pump, water is raised by means of the centrifugal force given to the water in a curved vane or arm, proceeding from the vertical axis. The dynamic principles of this machine are the same as those of the reaction wheel (Art. 158); but they differ in their objects. In the latter machine, a fall of water gives a rotatory motion to a vertical axis, while in the former a rotatory motion is given to a vertical axis in order to elevate a column of water.

Let h be the height to which the water is raised, measured from the level of the water in the well to the centre of the orifice of discharge, v the velocity of discharge through the orifice, and V the velocity of the orifice in its circular path, as in Art. 158. Then the work due to the centrifugal force must equal the work of raising the water through the height h, increased by the work stored in the water at efflux; therefore

$$\frac{WV^2}{2g} = Wh + \frac{Wv^2}{2g};$$

$$\therefore v = \sqrt{V^2 - 2gh}, \qquad (1)$$

and $\quad v - V = \sqrt{V^2 - 2gh} - V$

[as in (2) of Art. 158].

Now the work applied per second to raise each lb. of water must equal the work in raising the water through the height h, increased by the work remaining in the water after leaving the machine. Hence

$$\text{applied work} = h + \frac{(v-V)^2}{2g}$$

$$= \frac{(V - \sqrt{V^2 - 2gh})V}{g}. \qquad (2)$$

The useful work is h foot-pounds per second; therefore

$$\text{efficiency} = \frac{gh}{(V - \sqrt{V^2 - 2gh})V} \qquad (3)$$

$$= 1 - \frac{gh}{2V^2} - \text{etc.,} \qquad (4)$$

which increases towards the limit 1 as V increases towards infinity. Neglecting friction, therefore, the maximum efficiency is reached when the pump has an infinitely great velocity of rotation, as in the case of the reaction wheel.

Cor.—When $V^2 = 2gh$, we have, from (3),
$$\text{efficiency} = 0.5.$$
When $V^2 = 4gh$, we have, from (3),
$$\text{efficiency} = \frac{1}{2(2-\sqrt{2})} = .85.$$
When $V^2 = 6gh$, we have, from (3),
$$\text{efficiency} = 0.9.$$

Hence, theoretically, *the centrifugal pump has a considerable efficiency when the velocity of rotation exceeds the velocity due to twice the height of the column of water raised.*

Sch.—Centrifugal pumps work to the best advantage only at the particular lift for which they are designed. When employed for variable lifts, as is constantly the case in practice, their efficiency is much reduced, and does not exceed .5, and is often much less.

The earliest idea of a centrifugal pump was to employ an inverted Barker's Mill, consisting of a central pipe dipping into water, connected with rotating arms placed at the level at which water is to be delivered. The first pump of this kind which attracted notice was one exhibited by Mr. Appold in 1851, and the special features of this pump have been retained in the best pumps since constructed. The experiments conducted at the Great Exhibition on Appold's Centrifugal Pump with curved arms, gave the maximum efficiency 0.68. But when the arms were straight and radial, the efficiency was as low as .24, showing the great advantage of having the curved form of the arms, which causes the water to be projected in a tangential direction.

160. Turbines.—A reaction wheel is defective in principle, because the water after delivery has a rotatory velocity, in consequence of which a large part of the head is wasted (Art. 158). To avoid this, it is necessary to employ a machine in which some rotatory velocity is given to the water before entrance, in order that it may be possible to discharge it with no velocity except that which is absolutely required to pass it through the machine. Such a machine is called a *Turbine*, and it is described as "outward flow," "inward flow," or "parallel flow," according as the water during its passage through the machine diverges from, converges to, or moves parallel to the axis of rotation.*

Turbines are wheels, generally of small size compared with water wheels, driven chiefly by the impulse of the water. The water is allowed, before entering the moving part of the turbine, to acquire a considerable velocity; during its action on the turbine this velocity is diminished, and the impulse due to the change of momentum drives the turbine.

Roughly speaking, the fluid acts in a water-pressure engine directly by its pressure; in a water wheel chiefly by its weight causing a pressure, but in part by its kinetic energy, and in a turbine chiefly by its kinetic energy, which again causes a pressure.†

In the outward and inward flow turbines, the water enters and leaves the turbine in directions normal to the axis of rotation, and the paths of the molecules lie exactly or nearly in planes normal to the axis of rotation. In outward-flow turbines the general direction of flow is away from the axis, and in inward-flow turbines towards the axis. In parallel-flow turbines, the water enters and leaves the turbine in a direction parallel to the axis of rotation, and the paths of the molecules lie on cylindrical surfaces concentric with that axis.

There are many forms of outward-flow turbines, of which the best known was invented by Fourneyron, and is commonly known by his name. The inward-flow was invented by Prof. Jas. Thomson.

---

* Cotterill's App. Mechs., p. 506.   † Ency. Brit., Vol. XII., p. 520.

The theory of turbines is too intricate a subject to be considered in this treatise. For a general classification of turbines, with descriptions, illustrations, and discussions of these machines, as well as for a further development of hydraulic machines in detail, the student is referred, among other treatises, to the following: Fairbairn's Millwork and Machinery, Colyer's Water-Pressure Machinery, Barrow's Hydraulic Manual, Glynn's Power of Water, Prof. Unwin's Hydraulics.

EXAMPLES.

1. In a hydrostatic bellows (Fig. 70), the tube A is $\frac{1}{8}$ of an inch in diameter, and the area DE is a circle, the diameter of which is a yard. Find the weight which can be supported by a pressure of 1 lb. on the water in A.
Ans. 82,944 lbs.

2. Describe the siphon and its action. What would be the effect of making a small aperture at the highest point of a siphon?

3. A prismatic bell is lowered until the surface of the water within is 66 feet below the outer surface; state approximately how much the air is compressed.
Ans. To $\frac{1}{3}$ of its original volume.

4. If a prismatic bell 10 feet high be sunk in sea water until the water rises half way up the bell, find how far the top of the bell must sink below the surface, the temperature remaining the same.

Assume the water barometer = 33 feet for sea water.
Ans. 28 feet.

5. In the position of the bell in Ex. 4, find how much air must be forced into it in order to keep the water down to a level of 2 feet from its bottom.

Ans. 0.72 W, where W is the weight of the air in the bell when at the surface.

EXAMPLES. 295

6. If a small hole be made in the top of a diving bell, will the water flow in or the air flow out?

7. If a cylindrical diving bell, height 5 feet, be let down till the depth of its top is 55 feet, find (1) the space occupied by the air, and (2) the volume of air that must be forced in to expel the water completely, the water barometer standing at 33 feet.

Ans. (1) 1.8 nearly; (2) $\frac{20}{33}$ths of the volume of the bell.

8. The weight of a diving bell is 1120 lbs., and the weight of the water it would contain is 672 lbs. Find the tension of the rope when the level of the water inside the bell is 17 feet below the surface ($h = 33$ feet).

Ans. 676.48 lbs.

9. A cylindrical diving bell of height a is sunk in water till it becomes half full. Show that the depth from the surface of the water to the top of the bell is $h - \frac{a}{2}$.

10. A cylindrical diving bell, of which the height inside is 8 ft., is sunk till its top is 70 feet below the surface of the water. Find the depth of the air space inside the bell ($h = 33$ feet). *Ans.* $2\frac{1}{3}$ feet.

11. (1) Describe the action of a common pump; (2) distinguish between a lifting pump and a forcing pump; (3) to what height could mercury be raised by a pump?

12. The length of the lower pipe of a common pump above the surface of the water is 10 feet, and the area of the upper pipe is 4 times that of the lower; prove that if at the end of the first stroke the water just rises into the upper pipe, the length of the stroke must be 3 feet 7 inches very nearly ($h = 33$ feet).

13. If the diameter of the piston be 3 inches, and if the height of the water in the pump be 20 feet above the well, what is the pressure on the piston? *Ans.* 61.2 lbs.

14. If the diameter be $3\frac{1}{2}$ inches, the height of the water in the pump 27 feet 5 inches, the lever handle 4 feet, and the distance from the fulcrum to the end of the piston rod 4 inches, find the force necessary to work the pump-handle.
Ans. $9\frac{1}{2}$ lbs.

15. The height of the column of water is 60 feet above the well, the piston has a diameter of 3 inches, the pump-handle is $3\frac{1}{4}$ feet from the fulcrum, and the distance of the fulcrum from the piston rod is $3\frac{1}{2}$ inches; find the force necessary to work the pump. *Ans.* 15.3 lbs.

16. If the height of the cistern above the well be 25 feet, the diameter of the piston 2 inches, and the leverage of the handle 12 : 1, find the force necessary to use in pumping.
Ans. 2.83 lbs.

17. If the height of the cistern be 42 feet, the diameter of the piston $4\frac{1}{2}$ inches, the length of the handle 49 inches, and the distance of the fulcrum from the piston rod $3\frac{1}{4}$ inches, find the force. *Ans.* 20.65 lbs.

18. The diameter of the piston of a lifting pump is 1 foot, the piston range is $2\frac{1}{2}$ feet, and it makes 8 strokes per minute; find the weight of water discharged per minute, supposing that the highest level of the piston range is less than 33 feet above the surface in the reservoir ($h = 33$ feet). *Ans.* 312.5π lbs., or about 983 lbs.

19. If in working the pump of Ex. 18, the lower level of the piston range be $31\frac{1}{4}$ feet above the surface in the reservoir, find the weight of water discharged per minute.
Ans. 187.5π lbs.

20. In a Bramah's press $FO = 1$ inch, $FH = 4$ inches, the diameter of $A = 4$ inches, and diameter of $C = \frac{1}{2}$ an inch; find the force on A produced by a force of 2 lbs. applied at H. *Ans.* 512 lbs.

21. In one of the Bramah presses used in raising the Britannia tube over the Menai Straits, the diameter of the piston C was 1 inch, that of A 20 inches; the force applied to C at each stroke was $2\frac{1}{2}$ tons; find the lifting force produced by the upward motion of A. *Ans.* 1000 tons.

22. If the receiver be 4 times as large as the barrel of an air-pump, find after how many strokes the density of the air is diminished one-half.

Ans. Early in the 4th stroke.

23. After a very great number of strokes of the piston of an air-pump the mercury stands at 30 inches in the barometer-gauge, the capacity of the barrel being one-third that of the receiver, prove that after 3 strokes the height of the mercury is very nearly $12\frac{2}{3}$ inches.

24. A fine tube of glass, closed at the upper end, is inverted, and its open end is immersed in a basin of mercury, within the receiver of a condenser; the length of the tube is 15 inches, and it is observed that after 3 descents of the piston the mercury has risen 5 inches; how far will it have risen after 4 descents?

Ans. The ascent x is given by the equation $\frac{15}{15-x} + \frac{x}{h}$
$= \frac{4}{3} + \frac{20}{3h}$. If $h = 30$, $x = 6.1$ nearly.

25. If $A = 3B$ (Art. 151), find the elastic force of the air in the receiver after the 5th, 10th, 15th, and 20th strokes, the height of the barometer being 30 inches.

Ans. 7.119 ins.; 1.689 ins.; 0.401 ins.; 0.095 ins.

26. In the same pump, the barometer standing at 30, find the number of strokes, (1) when the mercury in the gauge rises to 25 inches, and (2) when the rarefaction is $1 \div 100$. *Ans.* (1) 6.2; (2) 16.

27. If a hemispherical diving bell be sunk in water until the surface of the water inside the bell bisects its vertical

radius, find the depth of the bell, supposing the atmospheric pressure to be 14.28 lbs. to the square inch ($h = 34$).
Ans. From surface to surface 73.3 feet.

28. There is a pump lifting water 29 feet high, the diameter of its piston is 1 foot, the play of the piston is 3 feet, and the pump makes 10 strokes per minute; (1) how many gallons of water will be discharged per minute, and (2) what is the pressure on the piston?
Ans. (1) 147 gals.; (2) 1420 lbs.

29. Water flowing through a trough, 2 feet wide and 1 foot deep, with a velocity of 10 feet per second falls upon an overshot wheel 50 feet in diameter. Find (1) the part of the fall operating by impulse; (2) the maximum useful work of the wheel, the efficiency being 0.70; and (3) the number of revolutions the wheel makes per hour when doing maximum work.
Ans. (1) 1.55 feet; (2) 43076.25 ft.-lbs. per sec.; (3) 114.59.

30. Water is furnished to a breast-wheel at the rate of 20 cubic feet per second with a velocity of 8 feet. The fall is 20 feet and the efficiency 0.75. What is the useful work done by the wheel when the periphery has a velocity of 3, 4, and 5 feet per second respectively? (See Sch. 1, Art. 155).
Ans. 19185.94; 19215.94; and 19185.9 ft.-lbs. per sec. respectively.

31. What is the useful work done by an undershot wheel, 40 feet in diameter, making 120 revolutions per hour, the velocity of the water being 20 feet per second and the area of the vanes being $1\frac{1}{2}$ square feet?
Ans. 1910 ft.-lbs. per second.

32. What is the efficiency of a reaction wheel when the water having a head of 16 feet, issues from the orifices with a velocity of 45 feet per second? *Ans.* 0.8254.

www.ingramcontent.com/pod-product-compliance
Lightning Source LLC
Chambersburg PA
CBHW022051230426
43672CB00008B/1138